MATH/STAT
LIBRARY

The Global Dynamics of Cellular Automata

The Global Dynamics of Cellular Automata

AN ATLAS OF BASIN OF ATTRACTION FIELDS OF
ONE-DIMENSIONAL CELLULAR AUTOMATA

Andrew Wuensche
Santa Fe Institute
Santa Fe, New Mexico, U.S.A.

Mike Lesser
International Ecotechnology Research Centre
Cranfield Institute of Technology
Bedford, MK43 0AL, United Kingdom

Reference Volume I

Santa Fe Institute
Studies in the Sciences of Complexity

ADDISON-WESLEY PUBLISHING COMPANY
The Advanced Book Program

Reading, Massachusetts   Menlo Park, California   New York
Don Mills, Ontario   Wokingham, England   Amsterdam   Bonn
Sydney   Singapore   Tokyo   Madrid   San Juan
Paris   Seoul   Milan   Mexico City   Taipei

Publisher: *David Miller*
Production Manager: *Michael Cirone*

Director of Publications, Santa Fe Institute: *Ronda K. Butler-Villa*
Publications Assistant, Santa Fe Institute: *Della L. Ulibarri*

The front cover shows a basin of attraction consisting of 8580 global states (about 26% of state space) converging onto an attractor cycle with period 120. The system parameters are $n = 5$, rule 54461424, $L = 15$, seed singleton.

The title page shows a basin of attraction which may be seen in the context of its basin of attraction field on page 203. The system parameters are $n = 5$, code 53, $L = 15$, seed singleton.

These and all similar graphics are examples of screen or printer output from the software included with the book.

Library of Congress Cataloging-in-Publication Data

Wuensche, Andrew.
    The global dynamics of cellular automata : an atlas of basin of attraction fields of one-dimensional cellular automata / Andrew Wuensche, Mike Lesser.
       p. cm. — (Santa Fe Institute studies in the sciences of complexity. Reference volumes ; v. 1)
    Includes bibliographical references (p.) and index.
    ISBN 0-201-55740-1
    1. Cellular automata. 2. State-space methods. 3. Differentiable dynamical systems.   I. Lesser, Mike. II. Title. III. Series.
QA267.5.C45W83   1992   511.3 — dc20   92-17543

This volume was typeset using TeXtures on a Macintosh II computer. Camera-ready output from a NEC Silentwriter 2 printer.

Copyright © 1992 by Addison-Wesley Publishing Company, The Advanced Book Program, Jacob Way, Reading, MA 01867.

All rights reserved. No part of this publication may be reproduced, stored in a retrieval system, or transmitted in any form or by any means, electronic, mechanical, photocopying, recording, or otherwise, without the prior written permission of the publisher. Printed in the United States of America. Published simultaneously in Canada.

Jacket design by Hannus Design Associates

1 2 3 4 5 6 7 8 9-AL-95949392
First printing, July 1992

About the Santa Fe Institute

The *Santa Fe Institute* (SFI) is a multidisciplinary graduate research and teaching institution formed to nurture research on complex systems and their simpler elements. A private, independent institution, SFI was founded in 1984. Its primary concern is to focus the tools of traditional scientific disciplines and emerging new computer resources on the problems and opportunities that are involved in the multidisciplinary study of complex systems—those fundamental processes that shape almost every aspect of human life. Understanding complex systems is critical to realizing the full potential of science, and may be expected to yield enormous intellectual and practical benefits.

All titles from the *Santa Fe Institute Studies in the Sciences of Complexity* series will carry the SFI logo; this imprint is based on a Mimbres pottery design (circa A.D. 950–1150) and was drawn by Betsy Jones.

Santa Fe Institute Editorial Board
June 1991

Dr. L. M. Simmons, Jr., *Chair*
Vice President, Academic Affairs, Santa Fe Institute

Professor Kenneth J. Arrow
Department of Economics, Stanford University

Professor W. Brian Arthur
Dean & Virginia Morrison Professor of Population Studies and Economics, Food Research Institute, Stanford University

Professor Michele Boldrin
MEDS, Northwestern University

Dr. David K. Campbell
Director, Center for Nonlinear Studies, Los Alamos National Laboratory

Dr. George A. Cowan
Visiting Researcher, Santa Fe Institute and Senior Fellow Emeritus, Los Alamos National Laboratory

Professor Marcus W. Feldman
Director, Institute for Population & Resource Studies, Stanford University

Professor Murray Gell-Mann
Division of Physics & Astronomy, California Institute of Technology

Professor John H. Holland
Division of Computer Science & Engineering, University of Michigan

Dr. Bela Julesz
Head, Visual Perception Research, AT& T Bell Laboratories

Professor Stuart Kauffman
School of Medicine, University of Pennsylvania

Dr. Edward A. Knapp
President, Santa Fe Institute

Professor Harold Morowitz, University Professor, George Mason University

Dr. Alan S. Perelson
Theoretical Division, Los Alamos National Laboratory

Professor David Pines
Department of Physics, University of Illinois

Professor Harry L. Swinney
Department of Physics, University of Texas

Santa Fe Institute Studies in the Sciences of Complexity

PROCEEDINGS VOLUMES

Volume	Editor	Title
I	David Pines	Emerging Syntheses in Science, 1987
II	Alan S. Perelson	Theoretical Immunology, Part One, 1988
III	Alan S. Perelson	Theoretical Immunology, Part Two, 1988
IV	Gary D. Doolen et al.	Lattice Gas Methods for Partial Differential Equations, 1989
V	Philip W. Anderson, Kenneth J. Arrow, & David Pines	The Economy as an Evolving Complex System, 1988
VI	Christopher G. Langton	Artificial Life: Proceedings of an Interdisciplinary Workshop on the Synthesis and Simulation of Living Systems, 1988
VII	George I. Bell & Thomas G. Marr	Computers and DNA, 1989
VIII	Wojciech H. Zurek	Complexity, Entropy, and the Physics of Information, 1990
IX	Alan S. Perelson & Stuart A. Kauffman	Molecular Evolution on Rugged Landscapes: Proteins, RNA and the Immune System, 1990
X	Christopher G. Langton et al.	Artificial Life II: Proceedings of the Second Interdisciplinary Workshop on the Synthesis and Simulation of Living Systems, 1991
XI	John A. Hawkins & Murray Gell-Mann	Evolution of Human Languages, 1992

LECTURES VOLUMES

Volume	Editor	Title
I	Daniel L. Stein	Lectures in the Sciences of Complexity, 1989
II	Erica Jen	1989 Lectures in Complex Systems, 1990
III	Daniel L. Stein & Lynn Nadel	1990 Lectures in Complex Systems, 1991

LECTURE NOTES VOLUMES

Volume	Author	Title
I	John Hertz, Anders Krogh, & Richard G. Palmer	Introduction to the Theory of Neural Computation, 1990
II	Gérard Weisbuch	Complex Systems Dynamics, 1990

REFERENCE VOLUMES

Volume	Author	Title
I	Andrew Wuensche & Mike Lesser	The Global Dynamics of Cellular Automata: An Atlas of Basin of Attraction Fields of One-Dimensional Cellular Automata, 1992

To my wife Stephanie, and my daughters Silole and Alice.

— Andrew Wuensche

To the memory of the achievements and the tragic end of Alan Turing.

— Mike Lesser

Foreword

There are a wide variety of methods for representing the behavior of dynamical systems. Perhaps the most familiar representation method is the traditional time-series plot, in which some observable variable of the system (e.g., angular position) is plotted on the vertical axis, with time progressing to the right on the horizontal axis.

Such time-series plots trace the behavior of a system through time from a specific initial state. Thus, such plots represent the behavior of a system "localized" to a particular initial state, and are referred to as "local" representations of behavior. In order to get a feeling for the "global" behavior of a system, behavior independent of any particular initial state, one can collect an ensemble of such time-series plots, each rooted at a different initial state, and superimpose them together in the same plot. For certain systems, such ensembles of local representations can, in fact, lead to useful insights into the global dynamics of the system.

However, the "state-space" representation, introduced by Poincaré, provides a much clearer portrait of the global behavior of dynamical systems. In a state-space representation, the ensemble of all possible time series is captured in the notion of a vector-field on the state space: the "field of flow" imposed on the space of states by a particular dynamical rule. A great deal of insight can be gained into the behavior of dynamical systems by understanding specific behaviors in terms of the topological properties of their associated trajectories in state space.

Although much of the work in the state-space analysis of dynamical systems has been carried out in the context of continuous state spaces, many of the concepts and methods carry over to discrete state spaces. In a discrete state space, the flow field can be seen to be a graph, in which the states are the nodes and the "flow" is captured by the edges linking the nodes. Just as one may have fixed points, limit cycles, and chaotic attractors in continuous flow fields, one may have fixed points, cycles, and infinite chains in graphs (in the latter case, of course, the state space must be infinite). Concepts such as the degree of spreading of a local patch of the flow field in continuous state spaces have their analogs in the degree of convergence—or "in-degree"—of a node in the flow graph in discrete state spaces.

The study of Cellular Automata (CA) has proven to be a particularly rewarding vehicle for gaining insights into the behaviors and peculiarities of discrete dynamical systems. However, a good deal of the previous analysis of CA has been carried out via the equivalent of time-series perspective, in which various properties of the space-time diagrams of the evolution of CA's from specific initial states are investigated.

This Atlas presents a comprehensive overview and analysis of CA from the state-space perspective. Although explicitly treating CA, many of the observations and results derived here depend only on properties of the flow graphs themselves, and consequently should be equally valid when applied to the flow graphs for other discrete dynamical systems.

This Atlas, together with the associated program for generating and analyzing flow graphs, should prove to be an invaluable tool for pursuing, in the context of discrete dynamical systems, the kinds of insights that can only be obtained from a global perspective.

Christopher Langton

Santa Fe, New Mexico
November 21, 1991

Preface

The study of the dynamical behavior of cellular automata (CA) has become a significant area of experimental mathematics in recent years. CA provide a mathematically rigorous framework for a class of discrete dynamical systems that allow complex, unpredictable behaviour to emerge from the deterministic local interactions of many simple components acting in parallel.

Such *emergent* behavior in complex systems, relying on *distributed* rather than centralised control, has become the accepted paradigm in the attempt to understand biology in terms of physics (and vice versa?), encompassing such great enigmas as the phenomena of life and the functioning of the brain. Rather than confronting these questions head on, an alternative strategy is to pose the more modest question: how does emergent behaviour arise in CA, one of the simplest examples of a complex system.

In this book we examine CA behaviour in the context of the global dynamics of the system, not only the unique trajectory of the system's future, but also the multiple merging trajectories that could have constituted the system's past.

In a CA, discrete values assigned to an array of sites change synchronously in discrete steps over time by the application of simple local rules. *Information structures*, consisting of propagating ensembles of values, may *emerge* within the array, and interact with each other and with other less active state configurations. Such emergent behaviour has lead to the notion of *computation emerging spontaneously* close to what may be a *phase transition* in CA rule space. Emergent behaviour in 2-D CA has given rise to the new field of *artificial life*.

In the simpler case of 1-D CA, a trace through time may be made which completely describes the CA's evolution from a given initial configuration. This is portrayed as rows of successive *global states* of the array, the *space-time pattern*. Space-time patterns represent a deterministic sequence of global states evolving along one particular path within a *basin of attraction*, familiar from continuous dynamical systems. In a finite array, the path inevitably leads to a state cycle. Other sequences of global states typically exist leading to the same state cycle. The set of all possible paths make up the basin of attraction. CA basins of attraction are thus composed of global states linked according to their evolutionary relationship, and will typically have a topology of branching trees rooted on attractor cycles.

Other separate basins of attraction typically exist within the set of all possible array configurations (*state space*). A CA will, in a sense, crystallise state space into a set of basins of attraction, known as the *basin of attraction field*. The basin of attraction field is a mathematical object which, if represented as a graph, is an explicit global portrait of a CA's entire repertoire of behaviour. It includes all possible space-time patterns.

The study of basin of attraction fields as a function of CA rule systems, and how the topology of the fields unfold for increasing array size, may lead to insights into CA behaviour, and thus to emergent behaviour in general. This book shows CA basin of attraction fields as computer graphics diagrams, so that these objects may be as easily accessible as space-time patterns in experimental mathematics.

Construction of basin of attraction fields poses the problem of finding the complete set of alternative global states that could have preceded a given global state, referred to as its *pre-images*. Solving this problem is recognised as being very difficult, other than by the explicit testing of the entire state space. Explicit testing becomes impractical in terms of computer time as the array size increases beyond modest values. Consequently, access to these objects has been limited.

This book introduces a *reverse algorithm* that directly computes the pre-images of a global state, resulting in an average computational performance that is many orders of magnitude faster than explicit testing. Two computer programs using the algorithm are described (and enclosed), to draw either basin of attraction fields or space-time patterns, for all 1-D, binary, *5-neighbour* CA rules, with *periodic boundary conditions*, and for the subsets of these rules, the *3-neighbour* rules, and the *5-neighbor totalistic* rules.

An atlas is presented (Appendix 2) showing the basin of attraction fields of all *3-neighbour* rules and all *5-neighbour* totalistic rules, produced using the program, for a range of array lengths. The atlas may be used as an aid to navigation in exploring the global dynamics of the 2^{32} rules in 5-neighbour rule space.

The book is divided into two parts. The first part (Chapters 1 through 4) gives the theoretical background and some implications of basin of attraction fields. The second part consists of appendices including the atlas and computer-program operating instructions.

Chapter 1 is an overview of the contents of the book.

Chapter 2 describes how CA global dynamics are represented by basin of attraction fields.

Chapter 3 looks in detail at CA architecture and rule systems, and the corresponding global dynamics. It is shown that *ordered architecture* and *periodic boundary conditions* impose restrictions on CA evolution in that *rotational symmetry* (and *bilateral symmetry* for *symmetrical rules*) are conserved. The rule numbering system and *equivalence classes* are reviewed. *Symmetry categories*, *rule clusters*, *limited pre-image rules*, and the *reverse algorithm* are introduced. The Z parameter, which reflects the *degree of preimaging*, or the convergence of dynamical flow in state space, is introduced.

Chapter 4 looks briefly at some implications of the above on current perceptions of the structure of *rule space*. The Z parameter is suggested as the mechanism underlying the λ *parameter*. A relationship between the Z parameter, basin field topology, and rule behaviour classes, based on the atlas, is proposed.

The idea of the rule table as *genotype* and the basin of attraction field as *phenotype* is examined. *Mutating* the rule table is found to result in *mutant* basin field topologies. Examples of sets of mutants are presented in Appendix 3.

We hope that the atlas of basin of attraction fields, and the program for exploring further into rule space, will provide new opportunities for CA research.

Acknowledgments

We are grateful to Grant Warrell for his suggestion (see page 55).

We are grateful to Bernardo Huberman for encouragement at a very early stage of the research.

We are grateful to Crayton Walker, Erica Jen, Wentian Li, and Mats Nordahl for discussions and comments.

Special thanks are extended to Chris Langton for discussions and suggestions, for his critical editing of the manuscript, and for supporting and promoting the project.

Thanks to Stuart Kauffman and Peter Allen for their invaluable encouragement and support.

Many thanks to Mike Simmons and to all the staff at the Santa Fe Institute for making us welcome, and for providing space and equipment on our visits to the institute.

We would like to especially thank Ronda Butler-Villa and Della Ulibarri for transforming a difficult manuscript into the present volume.

Table of Contents

	Foreword by Christopher Langton	xiii
	Preface	xv
ONE	Overview	1
TWO	Cellular Automata and the Basin of Attraction Field	5
	2.1 Cellular Automata	5
	2.2 The Basin of Attraction Field	8
THREE	The Transition Function and Global Dynamics	15
	3.1 General CA Parameters	15
	3.2 Rotation Symmetry	16
	3.3 Rule Clusters	18
	3.4 Limited Pre-image Rules	27
	3.5 The Reverse Algorithm	33
	3.6 The Z Parameter	39
FOUR	Implications of Basin of Attraction Fields	49
	4.1 Basin Field Topology and Rule Space	49
	4.2 Mutation	55
	4.3 Conclusion	58
APPENDIX 1	The Atlas Program	61
APPENDIX 2	Atlas of Basin of Attraction Fields	81
APPENDIX 3	Mutants	225
APPENDIX 4	The Rule-Space Matrix, $n=3$ Rules	235
	References	243
	Index	245

Overview
ONE

CHAPTER ONE: Overview

Outlines what appears in this volume.

CHAPTER TWO: Cellular Automata and the Basin of Attraction Field

2.1 Cellular Automata

Briefly reviews CA architecture, space-time patterns, and CA dynamics.

2.2 Basin of Attraction Fields

Describes how CA global dynamics may be represented by the basin of attraction field. Shows how state transition graphs and basin of attraction fields are constructed and depicted, given a *reverse algorithm* for directly computing *pre-images*. A program that draws basin of attraction fields, providing immediate access to these objects, is introduced. The scope of the program and Atlas are described. The significance of basin of attraction fields is briefly discussed.

CHAPTER THREE: The Transition Function and Global Dynamics

Looks in detail at the relationship between the structure of the CA transition function and the corresponding CA global dynamics.

3.1 General CA Parameters

Reviews general rule parameters for 1-D CA, and narrows them down to a system with a *local rule* and *wiring diagram*, and periodic boundary conditions.

3.2 Rotation Symmetry

Shows that such rules have the property that *rotation symmetry* cannot decrease, and must remain constant inside the attractor cycle. This restriction of behaviour explains the sensitivity of the basin field topology to the number theoretic properties of the array size, apparent from the atlas.

3.3 Rule Clusters

Reviews the rule numbering system[33,34,42] and *equivalence classes*[28,29,42] and introduces the concept of rule *symmetry categories* and *rule clusters*.[42] It is shown that symmetric rules conserve *bilateral symmetry*.

3.4 Limited Pre-image Rules

Introduces the *limited pre-image rules*, which are a subset of rules with a special *rule table* structure, whereby the number of pre-images to any global state has a fixed upper limit irrespective of array size.[42] The rule table structure of the limited pre-image rules provides the basis for the reverse algorithm.

3.5 The Reverse Algorithm

Explains the detailed implementation of the reverse algorithm, starting with the limited pre-image rules, and extending to all 1-D local rules in general.

3.6 The Z Parameter

Introduces a parameter Z, that reflects the *degree of pre-imaging* typical of a given rule, according to the distribution of values in the rule table. The Z parameter measures the probability that the *next cell* in a *partial pre-image* that is being computed by the reverse algorithm, has a unique value.

CHAPTER FOUR: Implications of Basin of Attraction Fields

Looks briefly at some implications of basin of attraction fields.

4.1 Basin Field Topology and Rule Space

Discusses the implications of basin fields on the current perception of the structure of CA rule space. The degree of pre-imaging is suggested as a determinant of basin field topology, and thus of rule behaviour classes. The Z parameter is suggested as the mechanism underlying Langton's λ parameter.[16,17]

4.2 Mutation

Looks at the effect of mutating the rule table by a small Hamming distance. It is shown that this generally results in related basin structure. Mating rules by combining 1/2 of the rule table of two related rules to create an offspring rule is also investigated. Examples of sets of mutants are shown in Appendix 4.

4.3 Conclusion

APPENDICES

Appendix 1: The Atlas Program—Operating Instructions

Contains the operation instructions (and graphic conventions) for two programs, *Atlas1* and *Space1*, for drawing either basin of attraction fields, or space-time patterns.

Atlas1 draws either the entire basin of attraction field or just a single basin, for any rule in the set of 5-neighbour rules, for a range of array length, L. This includes the 5-neighbour *totalistic codes*[34] and the 3-neighbour rules, also called *elementary* rules.[33,39] The basin field version of the program draws a graphic image of all rotation inequivalent basins in state space, up to $L = 18$. A single basin may be drawn up to $L = 31$, from any seed state that forms part of the basin. Selected data may be optionally printed as basin fields are drawn. The colour plates presented elsewhere in this volume are examples of screen output.

Space1 draws space-time patterns in various graphic formats for an array length L, up to $L = 640$.

The program on diskette is enclosed inside the back cover.

Appendix 2: Atlas of Basin of Attraction Fields

The Atlas consists of two parts.

Part 1 presents the basin fields for all 88 *equivalence classes* of the 3-neighbour rules for $L = 1$ to 15.

Part 2 presents the basin fields for all 36 equivalence classes of the 5-neighbour totalistic codes for $L = 3$ to 16.

There is an index to rules and codes at the beginning of each section. *Complementary* rules and codes are shown on facing pages. Selected data is also presented concerning each basin field. The key to a typical Atlas page layout is shown at the start of Appendix 2.

Appendix 3: Mutants
A 5-neighbour rule table has 32 mutations separated from it by one bit. Sets of mutant basins of attraction, and entire fields, derived from the source rule and its 32 one-bit mutants are illustrated. Sets of mutants that diverge progressively from the source rule are also illustrated.

Appendix 4: The Rule-Space Matrix
Describes a *rule-space matrix* for the 3-neighbour (elementary) rules. Manipulations of the matrix simulate equivalence and cluster transformations described in Section 3.3.

TWO

Cellular Automata and the Basin of Attraction Field

2.1 Cellular Automata

A cellular automaton (CA) is a discrete dynamical system which evolves by the iteration of a simple deterministic rule; as in any dynamical system, the system's variables change as a function of their current values.

An alternative approach is to view a CA as a parallel processing computer[34] where the data is considered to be the initial CA configuration.

Yet a third approach is that a CA is a "logical universe...with its own local physics."[16] Such a CA universe, in spite of its mathematically simple construction, seems to be capable of supporting complex emergent behaviour.

2.1.1 CA Architecture

A CA is constructed as follows: Time is discrete and progresses in steps. A D-dimensional, potentially infinite space is partitioned into discrete "cells" (the CA array or lattice) according to a given geometry. Boundary conditions may be set to define a finite space. Each cell has one attribute (the cell's value) from a limited range of attributes, which may be labelled by an integer. The pattern of values across the whole array is the CA global state at a given time.

Any pattern may be set as an initial condition at time t_0. Each cell of the array simultaneously has its value updated to evolve a new global state at time t_1. The new value of any given cell (the target cell) at t_1 is a function of the values and locations of a set of cells (the neighbourhood) at t_0, typically situated locally in relation to the target cell (see Fig. 2.1). The neighbourhood may be defined by a neighbourhood template or wiring diagram. The CA evolves through a succession of global states (its *trajectory*) by the iteration of this global updating procedure (the *transition function*). Provided that the transition function is constant and the system is closed to noise (the updating is error free), then the evolution of the CA from its initial global state is uniquely determined.

Two types of CA may be distinguished, both deterministic: The more general case may be described as having varying degrees of *disordered* architecture (non-local[21]), where the wiring diagram and/or function at each cell may differ, for example Walker's networks of Boolean functions[27-32] and Kauffman's random Boolean networks.[14,15]

CA with *ordered* architecture are a special case, where the wiring diagram and function are the same over the entire array. In addition, the ordered wiring may be confined to a *local neighbourhood*, an uninterrupted zone of cells typically centred on the target cell. CA of this type will be referred to as having *local architecture*, for example, the architecture of Wolfram's "elementary rules."[33-40] This paper is relevant to deterministic CA in general, however it deals mainly with the simplest possible local CA architecture.

Von Neumann first proposed CA to model self-reproduction.[3,26] His relatively complicated CA architectures were local and two-dimensional with 29 cell values. The tendency since then has been to find simpler architecture that could nonetheless support complex emergent behaviour. For example Conways' "game of life"[2] is a 2-D local

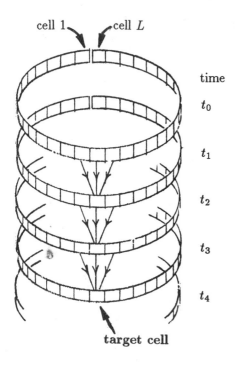

FIGURE 2.1 1-D, local binary CA with periodic boundary conditions, neighbourhood 3 (elementary rules), array length L.

CA with an orthogonal toroidal array, a 9-cell neighbourhood (the target cell and its eight nearest neighbours), and two cell values (binary value range).

2.1.2 Local 1-D Binary CA with Periodic Boundary Conditions

The simplest local CA architecture that we will investigate comprises a 1-D array of a small number of cells, a binary value range and a small local neighbourhood. The array is arranged in a circle; such a circular array is said to have *periodic boundary conditions*. Evolution of the CA may be represented as a sequence of global states on a cylinder, summed up by Fig. 2.1.

2.1.3 Space-Time Patterns

If the cylinder is split between cells 1 and L, and flattened out, it can be represented in 2-D as a space-time pattern, with space running across and time running down. (The space-time pattern of a 2-D CA could be represented in 3-D, for example[24]). Figure 2.2 shows examples of space-time patterns for various 1-D, local, binary, 5-neighbour rules. The rule numbers are indicated (see chapter 3, section 3.3.9).

Space-time patterns, which represent CA trajectories from given initial global states, have been the focus of statistical analysis and classification, and have been extensively illustrated in the literature, for instance.[17,18,33,34,39] Given the same CA architecture, different rules produce characteristic space-time patterns. For a given rule, patterns from different random initial global states are clearly recognisable as being similar by the human observer. Space-time patterns, in very broad terms, are said to display behaviour that is either static, periodic, complex (with interacting emergent structures), or chaotic.[17,34] CA rule classification schemes have been made on the basis of such space-time pattern phenomenology.

2.1 Cellular Automata

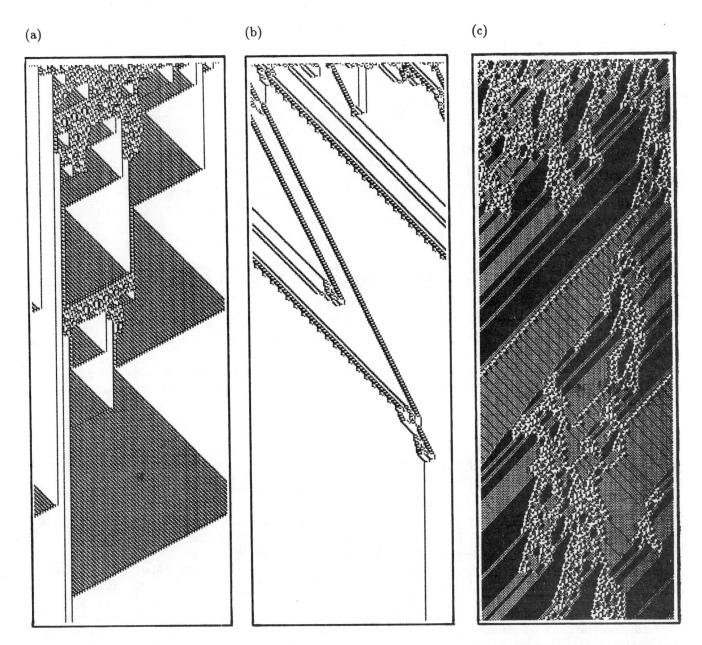

FIGURE 2.2 1-D, binary, local CA space-time patterns for 5-neighbour rules, with periodic boundary conditions. Array size 150, 420 time-steps from a random initial state. Rule numbers are (a) 3112581872; (b) 2334561936; and (c) 3583552890.

2.1.4 CA Dynamics

State space (also called *phase space*) is the set of all possible CA global states. In a finite CA, state space is finite; thus, any trajectory must eventually encounter a repeat of a global state that occurred at an earlier time. Because the system is deterministic, the trajectory will become trapped in this repeating sequence of states, a *cyclic attractor*, with a specific period of 1 or more.

States are either part of the attractor or belong to a *transient*, a sequence of states leading to the attractor. If transients exist, there must be states at their extremities (*garden-of-Eden states*), unreachable by evolution from any other state. The set of all possible transients leading to an attractor, plus the attractor itself, is the *basin of*

8   TWO Cellular Automata and the Basin of Attraction Field

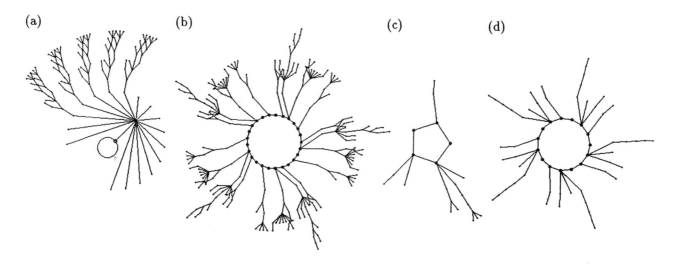

FIGURE 2.3 A basin of attraction field, local 3-neighbour rule 193, L = 10. The number of basins of each type is (a) 1, (b) 2, (c) 10, and (d) 2.

attraction of that attractor. State space is populated by one or more basins of attraction. These basins of attraction constitute the dynamical flow imposed on state space by the CA transition function.

A portrait of this global behaviour is the *basin of attraction field*, a discrete analogue of the familiar basin of attraction field found in the phase space of a continuous dynamical system, known as the system's *phase portrait*.

2.2 The Basin of Attraction Field

The *basin (of attraction) field* of a finite CA is the set of *basins of attraction* into which all possible states and trajectories will be organised by the cellular automaton transition function. The topology, or structure, of a single basin of attraction may be described by a diagram, the *state transition graph*. The set of graphs making up the field specifies the global behaviour of the system. Various other names have been used: state transition fragment,[39] contraction map,[7] topology of behaviour space,[29] and network of attraction.[42]

The notion of basin fields was proposed by Walker,[27] and examples[1] have been given by Martin et al.,[22] Pitsianis et al.,[25] Wolfram,[39,40] Feldberg and Rasmussen,[5] and by the authors[42] in an earlier edition of this atlas.

An example of a basin of attraction field is shown in Fig. 2.3 for the local, binary, 3-neighbour rule 193 (see chapter 3, section 3.3). The array length, L, equals 10, so that state space consists of $2^{10} = 1024$ global states. The CA transition function connects these states into a set of basins, the basin (of attraction) field. In this case there are four different types of basins in the field, some of which occur more than once. The number of each type is indicated.

2.2.1 The State Transition Graph

A state transition graph links up all the states belonging to a single basin of attraction according to their specific evolutionary location; this will typically have a topology of *trees rooted on attractor cycles*.[22] Global states are represented by *nodes* which are linked by *directed arcs*. Each node will have zero or more incoming arcs from nodes at the previous time step (*pre-images*), but because the system is deterministic, exactly one outgoing arc

---

[1]Martin et al.[22] have shown fields for rules 90 and 18. Pitsianis et al.[25] have shown fields for rule 90, with "null" boundary conditions. Wolfram has presented a table of sample basins for the 3-neighbour ($n = 3$) rules,[39] and the field for rule 30.[40] Feldberg and Rasmussen[5] have shown the field for the $n = 3$ rules for $L = 12$. In an earlier paper,[40] the authors produced an atlas of the fields for the $n = 3$ rules, and for totalistic codes 20 and 52.

2.2 The Basin of Attraction Field

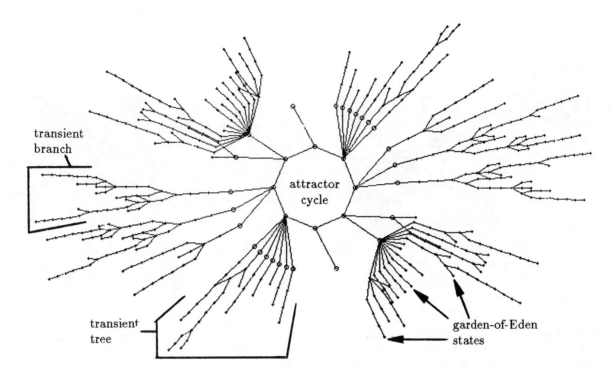

FIGURE 2.4 A state transition graph—basin of attraction (5-neighbour totalistic code 10, $L = 16$). Evolution proceeds inwards from garden-of-Eden states to the attractor, then clockwise. The graphic conventions are set out in Appendix 1.

("out degree") to a single node (the *successor state*), at the next time step. Nodes with no incoming arcs represent garden-of-Eden states. The number of incoming arcs is referred to as the *degree of pre-imaging* ("in degree").

For a given set of CA parameters, state space will, in a sense, crystallise into a set of one or more basins of attraction. The basin of attraction field is the set of state transition graphs representing all the basins in state space.

The make-up of a typical basin of attraction is illustrated by the state transition graph shown in Fig. 2.4 (it is part of the basin field shown in Fig. 2.6). In our graphic convention (see Appendix 2), the length of transition arcs decreases with distance away from the attractor, and the diameter of the graphic representation of the attractor asymptotically approaches an upper limit with increasing period, so that attractor cycles are drawn with approximately the same diameter irrespective of the number of nodes in the attractor. The forward direction of transitions is inward from garden-of-Eden states to the attractor, which is the only closed loop in the basin, and then clockwise around the attractor cycle.

Typically, the vast majority of nodes in a basin field, or a single basin of attraction, lie on *transient trees* outside the attractor cycle. A transient tree is the set of all paths from garden-of-Eden nodes leading to one node on the attractor cycle (an *attractor node*). A branch of the transient tree is termed a *transient branch*, and is the set of all paths from garden-of-Eden nodes leading to a state within a transient tree. A *transient* is one particular path from an arbitrary node in the transient tree leading to the attractor node. In all cases the attractor node itself is excluded from the definition.

2.2.2 Constructing the State Transition Graph

In our method of determining the topology of a single basin of attraction containing a particular global state, the attractor cycle is first isolated, then the topology of each transient tree is specified. This information is used to draw the state transition graph following a graphic convention described in detail in Appendix 1.

To isolate the attractor cycle, the CA is allowed to evolve forward in time from the selected initial state (the *seed*) until a repeat state is encountered. The number of steps to achieve this is the *disclosure length* (transient +

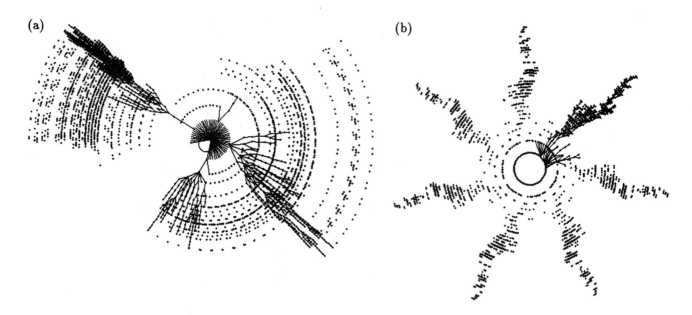

FIGURE 2.5 Examples of basins with (a) a *point attractor* (period 1, that cycles to itself), with suppressed equivalent transient branches [5-neighbour totalistic code 20, L=14, seed all 0s], and (b) a cyclic attractor (period 91), with suppressed equivalent transient trees [3-neighbour rule 193, L=14, singleton seed].

cycle period).[29] The sequence of states from the state that was repeated specifies the central attractor cycle of the basin.

The period of the attractor is generally between 1 and some simple multiple of L, the array length, but may diverge exponentially with L for *limited pre-image rules* (see chapter 3, section 3.4). The attractor period cannot, however, exceed $2^L - M$, where M is the number of states in state space made up of repeating segments on the circular array (see chapter 3, section 3.2).

Once the attractor cycle is known, it is drawn as a circle of nodes, evolving clockwise. To construct transient trees it is necessary to have a method of computing the complete set of pre-images of any node, in other words, to evolve the CA backwards in time. Given that such a *reverse algorithm* is available, the pre-images of an attractor state are computed, but the pre-image lying on the attractor cycle itself is deliberately excluded to prevent endlessly tracing pre-images "backwards" around the attractor cycle.

The reverse algorithm is reapplied repeatedly, to the pre-images of pre-images, until all the *garden-of-Eden states*,[12,14,22,23] those without pre-images, are reached. In this way the full description and topology of a *transient tree* is specified. The transient tree, if any, feeding into each state on the attractor cycle is derived in turn.

The construction of transient trees and branches is simplified by taking advantage of *shift invariance*.[40] There are global states that differ only by a rotation of the circular array. Such *rotation equivalent* states must have equivalent pre-images, rotated by the same amount, and must occupy equivalent positions in the same or an equivalent basin.

If the pre-images to a given state have been computed, the pre-images of its rotation equivalents are known, and by extension so is the entire transient tree (or transient branch), which does not need to be re-computed. If rotation equivalent states belong to separate basins, the basins will be equivalent, so only one example needs to be constructed.

Figure 2.5(a) shows a basin with equivalent transient branches from a point attractor (where all transients belong to a single transient tree). Figure 2.5(b) shows a basin with equivalent transient trees from an attractor cycle. In both cases only one example of a set of equivalent transient trees or branches has been drawn, and the remainder have been suppressed (apart from the point-attractor pre-images in Fig. 2.5(a)). Garden-of-Eden nodes, however, have been retained to indicate the footprints of the suppressed transients.

2.2 The Basin of Attraction Field

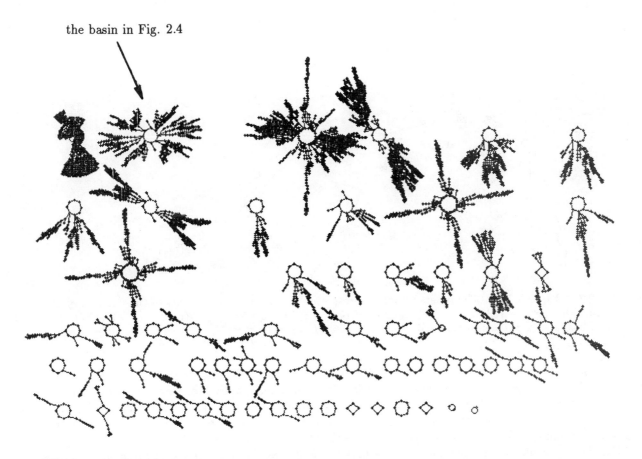

the basin in Fig. 2.4

FIGURE 2.6 The basin of attraction field of the 5-neighbour totalistic code 10, $L = 16$.

2.2.3 Constructing the Basin of Attraction Field

In our method of determining the set of basins that make up the basin of attraction field, states that have not been assigned to basins are used to seed new basins until no unassigned states remain.

Figure 2.6 illustrates the basin field of the 5-neighbour rule with totalistic code 10 (see chapter 3, section 3.3), for array length 16. There are 2^{16} global states that are organised by the rule into a basin field consisting of 769 separate basins, in 64 distinct rotation equivalent sets of basins. Only one example of each rotation equivalent set of basins is shown, as they have an identical topology. The basin indicated (one of a set of eight) is illustrated at a larger scale in Fig. 2.4.

In our graphic convention, global CA states in the state transition graph are normally represented as nodes of decreasing size with distance from the attractor. However, the actual binary state of the node, or its decimal equivalent, may be displayed at the node position. (The resolution is subject to the scale of the diagram; see Appendix 1). This allows the explicit global behaviour of CA to be fully described. Figure 2.7 illustrates a basin field with numbered nodes.

2.2.4 Pre-images

The discrete and finite nature of CA allow their basin of attraction fields to be depicted by a diagram that may, in principle, be totally explicit. However, the difficulty in computing the required pre-images for constructing basin of attraction fields has prevented easy access to these objects.

As described above, the basin field can be constructed if the pre-images to each state in state space are known. This information can be derived in general by exhaustive testing; to find the pre-images of a global state S_1, the

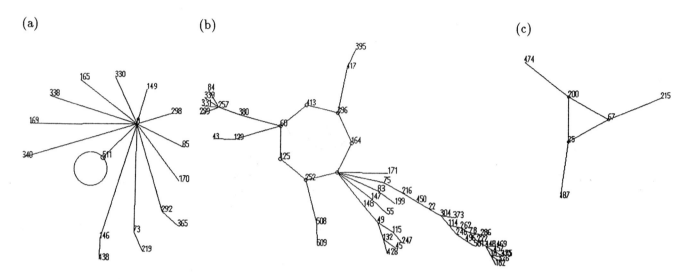

FIGURE 2.7 A basin field with numbered nodes representing the decimal equivalents of the CA binary global states. 3-neighbour rule 193, $L = 9$. The number of basins of each type is (a) 1, (b) 9, and (c) 3. Note that each of the set of basins of a particular type will consist of *rotation equivalent* states, and will have differently numbered nodes accordingly.

CA is evolved one step forward in time from all possible unallocated states. A one-step evolution that results in S_1 provides a pre-image of S_1. However, the number of states to be tested in a binary CA is equal to 2^L, where L is the array length. The corresponding exponential increase in the time required to compute pre-images for successively larger arrays rapidly makes their computation impractical.

Exhaustive testing has in general been the only method available for computing pre-images, thus effectively preventing a systematic view of basin fields. In Chapter 3 a computational shortcut, the *reverse algorithm*, is presented, which allows the pre-images of a CA state to be computed directly, with an average computational performance that is many orders of magnitude faster than explicit testing. This enables real-time computer generation of basin fields, thus providing a new opportunity for CA research.

2.2.5 The Program

A computer program that draws basin of attraction fields, or just a single basin, using the reverse algorithm, providing immediate access to these objects, is included with this book, together with a program that draws space-time patterns.

The programs work for any rule in the set of 1-D, binary, 5-neighbour rules. This includes the *totalistic codes*[34] and the 3-neighbour rules, also known as *elementary* rules.[33,39] The program operating instructions and the graphic conventions are described in detail in Appendix 1.

Space-time patterns are generated in various graphic formats for an array length L, up to $L = 640$. Space-time patterns can also be run "backwards," generating binary strings representing the pre-images of successive global states. Provided that the number of pre-images to a given state is within the limits of the computer system, this may be done for an array length up to $L = 80$.

The basin field version of the program draws a graphic image in real time of all rotation inequivalent basins in state space, for a selected range of L up to 18, together with relevant data. A single basin may be drawn up to $L = 31$, from any seed state that forms part of the basin. Alternatively, a fragment of a graph may be drawn representing only the transient branch leading to a given state. Selected data on each basin and on the basin field may be printed.

2.2 The Basin of Attraction Field

FIGURE 2.8 Example of a basin of the 3-neighbour rule 126, $L = 31$. The seed state is a single 0 followed by six copies of the string 01111.

2.2.6 The Atlas

The program has been used to produce an *atlas* of basin of attraction fields over a range of array lengths, presented in Appendix 2. The Atlas consists of two parts. Part 1 presents all 3-neighbour (elementary) rules. Part 2 presents all 5-neighbour totalistic codes. Selected data is also presented on each basin and basin field. There is an index to rules and codes at the beginning of each part.

A CA rule belongs to a set of up to four *equivalent* rules, the *equivalence class*, that differ only in that they have *negative* and/or *mirror image* space-time patterns, but which have identical basin field topology. Thus, only one rule representing each equivalence class is presented in the Atlas. Pairs of equivalence classes relate in that their rules have *complementary* rule tables, forming a *rule cluster*. The representative rules in the Atlas are presented according to their rule cluster, with complementary equivalence classes shown on facing pages (the relationships between rules are explained in detail in chapter 3, section 3.5)

The 3-neighbour rule clusters belong to one of three *symmetry categories—symmetric, semi-asymmetric*, and *fully asymmetric*—and are accordingly presented in three sections. The basin fields for the 88 *equivalence classes* of the 3-neighbour rules are shown for $L = 1$ to 15.

The 36 equivalence classes of the 5-neighbour totalistic codes (which belong only to symmetric clusters) are shown for $L = 3$ to 16.

The atlas may be used as an aid to navigation in exploring the basin field structure of any of the 2^{32} rules in 5-neighbour rule space, by mutating the rule table away from the subsets of rules with known basin fields, and moving into unknown territory.

2.2.7 Significance of Basin of Attraction Fields

The ability to represent basin of attraction fields may be significant in a number of areas such as CA theory, complex systems, dynamical systems, computational theory, artificial life, neural networks, and aspects of genetics.

Basin field topology represents a second order of complexity of CA behaviour, where the first order may be said to be the space-time patterns of particular trajectories. Easy access to a systematic presentation of basin fields, for a synoptic, qualitative, as well as explicit view of global behaviour, may provide insights into the dynamical theory of CA, and the structure of CA rule space. It may provide a useful analogue to the global behaviour of CA with more complex architecture, and dynamical systems in general.

The separate basins in a basin of attraction field classify state space. Attractors have been regarded as "memories,"[13,14,15] with implications for the mechanism underlying neural networks.

Basin fields may be of interest in genetics because mutations of the CA *rule table* by a small Hamming distance typically result in altered but related basin structures. Analogies have been made between a CA rule table and a DNA sequence.[20] Kauffman and others study CA with non-local architecture as models in biology and genetics.[14,15]

14   TWO Cellular Automata and the Basin of Attraction Field

Attractors have been interpreted as "cell types" in ontogeny. Evolution is said to occur in an optimal "fitness landscape" by mutation and selection of the CA "genotype" resulting in adapted dynamics or "phenotype."

A possible approach to basin of attraction fields is to see them as artificial morphologys, analogous to the "biomorphs" proposed by Dawkins.[4] Their explicit morphological form (including the global CA state at each node) is determined by the genetic code, the rule table; a mutation of the rule table results in a mutant morphology. In addition, the genetic code is implicit in the morphology because the rule table can be reconstructed from space-time patterns; this suggests possibilities for reproduction. Such a genotype-phenotype approach to CA may suggests applications in artificial life.

THREE

The Transition Function and Global Dynamics

3.1 General CA Parameters

In general, a cellular automaton is an array evolving by the iteration of a constant nonlinear function that independently determines the new value of each cell in the array according to the values of a predetermined subset of preceding cells. The array is restricted to one dimension and the definition is narrowed in stages. Fig. 3.1 illustrates a general space-time view of a CA adapted from the system proposed by Ashby.[1]

The cells of a finite array of size L, are restricted to k values (the *value range*), labelled for convenience $0, 1, 2, \ldots, k-1$. Each *target cell* at t_1 is wired to a subset of cells at t_0 and assigned a rule which acts on the set of cells specified by the *wiring diagram* (the order of input wires *is* significant), to determine the target cell's new value. The system may have *disordered architecture* in that both the wiring diagrams and the rules at each cell location may differ.

The updating of cell values is synchronous, and together specify the CA *transition function*, which is fixed over time. Note that *historical time reference* ("higher order in time" [34]), where the wiring diagram may extend to cells at t_{-1} and before, is excluded from this general scheme as it would result in qualitatively different behaviour from that described below.

Although these CA parameters are very broad, the following can be said about behaviour (and extended to two and higher dimensions), if the system is closed to external noise:

1. The global state of the array at t_0 has one and only one successor state at t_1.

2. The global state at t_0 may have 0, 1 or more pre-images at t_{-1}.

3. The total number of global states, the size of the state space, equals k^L.

4. Evolution will end at an attractor cycle, or at a point attractor which cycles on itself.

5. The transition function will organise all space-time trajectories into a basin of attraction field, which can, in principle, be represented by a set of state transition graphs, describing global behaviour.

6. The transition function will have equivalents: two equivalents by reflection (mirror-image space-time patterns, more equivalents for higher dimensions), and $k!$ equivalents by interchanging cell values.

The general scheme is complicated. This architecture may be drastically simplified without sacrificing complexity of behaviour. Walker's "networks of Boolean functions" [27-32] limit the value range to binary and the number of wiring inputs to 3. The target cell at t_1 is wired to itself at t_0, and the remaining two wires set at random, but with the same rule at each cell position. Walker has studied basin field structure by means of random sampling. He has sought to show the variation of basin topology for different rules and array sizes. In particular, he has investigated disclosure

15

THREE The Transition Function and Global Dynamics

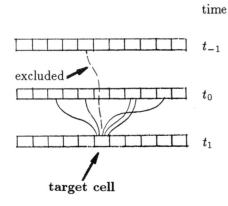

FIGURE 3.1 Disordered 1-D CA with arbitrary wiring from cells at t_0 to a target cell at t_1. The wiring and rule may differ at each target cell, but are constant over time. Wiring between t_1 and t_{-1} is excluded.

length (transient + cycle length) and the *dominance* of state space by the attractor,[32] which is the proportion of states in state space "drained" by an attractor.

Kauffman has investigated "random Boolean networks" as models of genetic systems and evolution.[14,15] In his CA architecture, both rule and wiring diagram (with typically only two wires) are set at random at each cell position. He has found that in spite of their random construction, "such systems can spontaneously crystallise enormously ordered dynamical behaviour." [14]

In such *disordered architecture* as described above, the question of defining the boundary conditions does not arise because each cell, including edge cells, has its wiring diagram individually specified.

3.1.1 Local Architecture

The situation is drastically simplified if the wiring diagram and rule are identical at each cell position. In the case of such *ordered architecture*, however, the wiring of cells at the edges of the array, the boundary conditions, needs to be defined. An additional simplification is *local architecture*, with nearest-neighbour wiring; the wiring is confined to an uninterrupted neighbourhood of cells, typically situated locally in relation to the target cell.

Wolfram has investigated local, binary CA, where every cell is wired uniformly to the three neighbouring cells centred on itself, (elementary rules).[33] Boundary conditions are specified as either infinite, fixed, or periodic (though fixing boundary conditions makes the system disordered, because the wiring is atypical at edge cell locations). Considerable descriptive and analytic work has focused on specific rules of this type.[8,9,20,22,33,39-42] Wolfram broadened his investigation by increasing the neighbourhood to 5 cells, and also increasing the value range to 3 for 3-neighbour rules.[34] Langton has investigated a wide variety of local CA architectures with various value ranges, and has suggested the existence of a *phase transition* in rule space.[16]

3.2 Rotation Symmetry

In this investigation, CA parameters are narrowed initially as follows. An ordered architecture, with periodic boundary conditions (edge cells wired to simulate a circular array). Each cell has the *same* (possibly randomly connected) wiring diagram. This architecture is summed up in Fig. 3.2.

Ordered architecture and periodic boundary conditions impose additional restrictions on dynamical behaviour. The *rotation symmetry*[42] (also known as *translational invariance*) is the maximum number of identical sequences of cell values, termed *repeating segments*, into which the circular array can be divided. The size of a repeating segment is the minimum number of cells through which the circular array may be rotated and still appear identical. The rotation symmetry of successive global states in CA evolution cannot decrease, and may only increase in a transient.

The potential for the array to acquire rotation symmetry depends on the *number theoretic properties* of the array length, L, which has general effects on the topology of basin of attraction fields and the length of attractor cycles. Martin et al.[22] drew attention to this feature in relation to the *additive rules*, and noted that it accounts

3.2 Rotation Symmetry

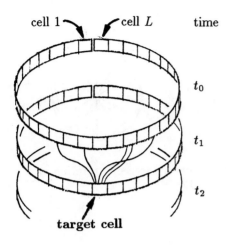

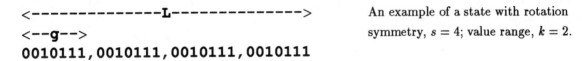

FIGURE 3.2 Ordered CA architecture with periodic boundary conditions. Each cell has the same rule/wiring.

for the irregular variation of attractor periods with L, and the "self-organising" behaviour of CA has been widely noted, for instance[22,34]

```
<---------------L--------------->
<--g-->
0010111, 0010111, 0010111, 0010111
```

An example of a state with rotation symmetry, $s = 4$; value range, $k = 2$.

Consider an array size, L, value range, k, with a total number of states, k^L. If the rotation symmetry is s, and the length of each repeating segment is g, then $s = L/g$. If $s = 1$, then the whole line is one indivisible segment, a *disordered state*, and $g = L$. If all cells have equal value, for instance $11111...$, then $s = L$ and $g = 1$. This is the highest possible degree of rotation symmetry, a *uniform state*. The number of uniform states is equal to the value range, k. If $s > 1$, then the state is a *segmented state*, where $g \leq L/2$.

Rotation equivalents are global states that differ only by a rotation of the circular array. Disordered states will have the maximum L rotation equivalents, whereas segmented states will have fewer g rotation equivalents.

There is no privileged segment or group of segments that can be treated differently from the rest in a circular array evolving under an ordered rule. Repeating segments are equivalent; therefore, evolution of all segments must be identical, and the following can be said in general about CA evolution:

1. Evolution cannot decrease s, as this would imply inequivalent evolution of segments, which is impossible if all segments must be treated identically.
2. On the attractor cycle, s must remain constant, because any increase would need to be balanced by a decrease as evolution returned around the cycle, and a decrease cannot occur as a consequence of 1.
3. In a transient, s will either remain constant or increase as a consequence of 1. As a consequence of 2, a trajectory where s increases must be a transient. As a consequence of 1, the location of states with greater s in the basin will be downstream, closer to the attractor. Transient evolution (see Fig. 3.3) can proceed in this order, but never in the opposite direction:

   disordered states ⟶ **segmented states**, s increasing ⟶ **uniform states**

4. s can only increase by equivalent subdivisions of g into equal parts; therefore, if L or g is prime, s can only increase by a jump to the uniform state.

These general restrictions on behaviour result in the following limits to the maximum attractor cycle period, P_{max}, and maximum disclosure length, D_{max}, in the basin of attraction field of any rule.

1. The k uniform states occur closer to the end of evolution than any other state, and cannot form part of a cycle with a non-uniform state; only a uniform state may be *downstream* of another uniform state; therefore, for any state where $g = 1$ ($s = L$), $P_{max} \leq k$ and $D_{max} \leq k$.

18 THREE The Transition Function and Global Dynamics

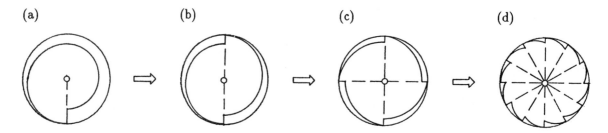

FIGURE 3.3 Possible transient evolution of a disordered state to states with an increasing degree of rotation symmetry. (a) $s = 1$, $g = L$. (b) $s = 2$, $g = L/2$. (c) $s = 4$, $g = L/4$. (d) $s = 12$, $g = L/12$.

2. For any state where $g > 1$, $P_{max} \leq k^g - k$ and $D_{max} \leq k^g$.
3. For any array of length L, $P_{max} \leq k^L - k$ and $D_{max} \leq k^L$.
4. For any state where $g > 1$, if m is the total of all states of length g which are made up of repeating segments, $P_{max} \leq k^g - m$ and $D_{max} \leq k^g$.
5. For any array of length L, if M is the total of all segmented states, $P_{max} \leq k^L - M$ and $D_{max} \leq k^L$.

The division of state space into segmented states M, and disordered states $k^L - M$, depends on the number of theoretic properties of the array length, L. The fact remains, however, that many basins are comprised only of disordered states; for prime L all are disordered apart from the uniform states. This leaves open the question of whether evolution among disordered states, and also the more general systems of Walker and Kauffman, follow parallel general principles

A possible approach is to consider *symmetric rules* (which include all *totalistic rules*) which conserve *bilateral symmetry* of the array. The degree of bilateral symmetry may increase in a transient, but must remain constant in the attractor. States with bilateral symmetry may be segmented, or disordered. For instance, the singleton state (a single 1 among 0s) is a disorderd state by our definition, but has bilateral symmetry (see section 3.3.6).

3.3 Rule Clusters

CA parameters may, finally, be restricted as follows: a binary value range ($k = 2$) and local architecture, thus nearest-neighbour wiring, and periodic boundary conditions (a circular array). For a neighbourhood size n, the number of different neighbourhoods equals 2^n, and the number of different rules equals 2^{2^n}.

3.3.1 Rule Numbering System, n=3 (Elementary Rules)

The $k = 2$, $n = 3$ rules have the form,

$$P_i^{t+1} = f(P_{i-1}^t, P_i^t, P_{i+1}^t)$$

where $P_i = 0$ or 1, i is the spatial position between 1 and L, and t is the time. For a circular array, length L, $P_1 = P_{L+1}$.

The system may be represented as follows:

```
t0     - - A B C -      B C - - - A      C - - - A B      The neightbourhood ABC at t0
t1     - - - T - -      T - - - - -      - - - - - T      defines the cell T at t1.
```

The $n = 3$ *rule table* with $2^3 = 8$ entries uniquely specifies each individual rule from a total of $2^{2^3} = 256$. Following Wolfram's convention[33] the rule table is ordered in descending values of the binary neighbourhood strings.

	111	110	101	100	011	010	001	000	neighbourhoods
Rule table..	T_7	T_6	T_5	T_4	T_3	T_2	T_1	T_0	outputs

3.3 Rule Clusters

If the rule table is regarded as a binary number, the rule number, R, is its decimal equivalent; thus, the rule table will range from 00000000 to 11111111, and R from 0 to 255.

3.3.2 Complementary Transformation, n=3

Every rule, R, has a distinct *complementary* rule, R_c, where each entry in the rule table is inverted, so that for a given input line, the next line generated by R_c will be the *negative* of the next line generated by R. R and R_c may have equivalent behaviour (see collapsed clusters below).[1] In any case, their behaviour will be closely related. This will be reflected in the basin field structure. Deterministic structure and symmetry category (see below) will be common to R and R_c. In general, pre-imaging, cycle periods, and transient lengths will be related.

The $n = 3$ rules can be listed in 128 complementary pairs.

$$\begin{array}{rcccccc}
\text{if } R = & 0, & 1, & 2, & \cdots, & 127 \\
& | & | & | & & | \\
R_c = & 255, & 254, & 253, & \cdots, & 128
\end{array}$$

There are two types of symmetry that relate pairs of neighbourhoods:

$$\begin{array}{lll}
\text{Complementary neighbourhood pairs} & \cdots \quad 111, 000 & T_7, T_0 \\
\text{(0s changed to 1s and vice versa)} & \quad 110, 001 & T_6, T_1 \\
& \quad 101, 010 & T_5, T_2 \\
& \quad 100, 011 & T_4, T_3 \\
\\
\text{Reflected neighbourhood pairs} & \cdots \quad 110, 011 & T_6, T_3 \\
\text{(mirror image)} & \quad 100, 001 & T_4, T_1
\end{array}$$

3.3.3 Negative Transformation, n=3

For any rule R and input line I, the CA will generate a space-time pattern P. There is a rule, R_n, that, given the negative input line, $\overline{I}$ will generate the negative space-time pattern, $\overline{P}$ (all cell values inverted).

Consider the rule table:

$$\begin{array}{lcccccccc}
& 111 & 110 & 101 & 100 & 011 & 010 & 001 & 000 & \text{neighbourhoods} \\
\text{Rule table..} & T_7 & T_6 & T_5 & T_4 & T_3 & T_2 & T_1 & T_0 & \text{outputs}
\end{array}$$

To find R_n suppose a *negative world* in which the rule table, including neighbourhoods, could be transformed by inverting all values into

$$\begin{array}{lcccccccc}
& 000 & 001 & 010 & 011 & 100 & 101 & 110 & 111 & \text{neighbourhoods} \\
\text{Rule table..} & \overline{T_7} & \overline{T_6} & \overline{T_5} & \overline{T_4} & \overline{T_3} & \overline{T_2} & \overline{T_1} & \overline{T_0} & \text{outputs}
\end{array}$$

The conventional order of neighbourhoods has been altered and must be restored, giving

$$\begin{array}{lcccccccc}
& 111 & 110 & 101 & 100 & 011 & 010 & 001 & 000 & \text{neighbourhoods} \\
\text{Rule table..} & \overline{T_0} & \overline{T_1} & \overline{T_2} & \overline{T_3} & \overline{T_4} & \overline{T_5} & \overline{T_6} & \overline{T_7} & \text{outputs}
\end{array}$$

---

[1] Although negative and complement may have the same meaning, changing 0s for 1s (permutation of values), we use *negative* (as in photographic negative) for transformed CA space-time patterns, and *complement* for transformed rule tables or neighbourhoods.

The transformed rule table is R_n. Thus the procedure to transform R to R_n is as follows in either order.
1. Take the complement of the rule table, R_c.
2. Swap the output of complementary neighbourhoods.

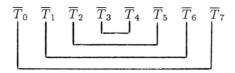

For example, rule 193-11000001 is transformed to rule 124-01111100.

3.3.4 Reflection Transformation, n=3

For any rule, R, input line I and space-time pattern P, there is a rule R_r that, given the reflected input line, I_r will generate the reflected (mirror-image) space-time pattern P_r.

Consider the rule table:

	111	110	101	100	011	010	001	000	neighbourhoods
Rule table..	T_7	T_6	T_5	T_4	T_3	T_2	T_1	T_0	outputs

To find R_r suppose a *mirror image world* in which the rule table, including neighbourhoods would be transformed by reflection into

	000	100	010	110	001	101	011	111	neighbourhoods
Ruletable..	T_0	T_1	T_2	T_3	T_4	T_5	T_6	T_7	outputs

The conventional order of neighbourhoods has been altered and must be restored, giving

	111	110	101	100	011	010	001	000	neighbourhoods
Rule table..	T_7	T_3	T_5	T_1	T_6	T_2	T_4	T_0	outputs

The transformed rule table is R_r. Note that only the outputs of asymmetric neighbourhoods are altered. Thus to transform R to R_r, swap the outputs of the two pairs of asymmetric reflected neighbourhoods.

$$T_7 \quad T_6 \quad T_5 \quad T_4 \quad T_3 \quad T_2 \quad T_1 \quad T_0 \quad \text{rule table}$$

For example, rule 193-11000001 is transformed to rule 137-10001001.

3.3.5 Symmetry Categories

The reflection transformation allows rules to be placed into one of three symmetry categories:
1. *symmetric rules* ($R = R_r$), if $T_6 = T_3$ and $T_4 = T_1$.
2. *semi-asymmetric rules*, if either $T_6 \neq T_3$ or $T_4 \neq T_1$, but not both.
3. *fully asymmetric rules*, if $T_6 \neq T_3$ and $T_4 \neq T_1$.

Symmetric rules have space-time patterns whose structures appear to have no bias to move left or right; for semi-asymmetric rules, there is a clear bias towards either the left or right; and for fully asymmetric rules, there is an intersecting bias towards both left and right.

3.3 Rule Clusters

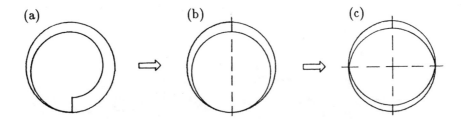

FIGURE 3.4 Possible transient evolution of a disordered state to a state with bilateral symmetry, bs (rotation symmetry, s). (a) $bs = 0$, $s = 0$. (b) $bs = 2$, $s = 0$. (c) $bs = 4$, $s = 2$.

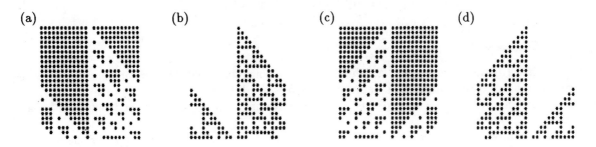

FIGURE 3.5 Equivalent space-time patterns, from equivalent rules. (a) rule R (193). (b) rule R_n (124). (c) rule R_r (137). (d) rule R_{nr} (110).

3.3.6 Bilateral Symmetry of Symmetric Rules

The space-time patterns of symmetric rules, given an input line with *bilateral symmetry* (bs), must conserve bilateral symmetry, because the rule acts equivalently on both sides of the axis of symmetry.

As with rotation symmetry (see section 3.2), the degree of bilateral symmetry cannot decrease; in an attractor, bilateral symmetry must remain constant, because any increase would have to be balanced by a decrease as evolution returned around the attractor cycle, and a decrease cannot occur. Bilateral symmetry may increase only in a transient (see Fig. 3.4).

The bilateral axis may divide a circular array, length L, as follows: if L is even, the axis may divide the array into two reflected halves, for instance $0001 - 1000$. If the maximum disclosure length is d_{max}, then $d_{max} \leq 2^{L/2}$. If L is odd, the axis may divide the array into two reflected halves bisecting one cell, for instance $000\underline{1}000$, then $d_{max} \leq 2^{((L-1)/2)+1}$. Alternatively, if L is even, there may be more than one axis of bilateral symmetry, for instance $\underline{00}01\underline{00}01$. Such an array has rotation symmetry as well as bilateral symmetry. The interaction between rotational and bilateral symmetry and consequences on behaviour is still unclear. All *totalistic rules* are symmetric (see 3.3.10).

3.3.7 Equivalence Classes

We have seen that a given rule, R, will have two equivalent rules, R_n and R_r. R will also have a third equivalent, R_{nr}, derived by performing the negative and reflection transformations successively, in either order. There will thus be a maximum of four rules grouped in an *equivalence class*; an example is shown in Fig. 3.5, based on rule 193.

The equivalent rules differ only in that they have negative and mirror-image space-time patterns; otherwise, CA behaviour is totally equivalent. There are 88 *equivalence classes* among the 256 $n = 3$ rules.[28,42] Basin field topology for the rules in each class will be identical, though the actual states will differ according to the transformations described.

THREE The Transition Function and Global Dynamics

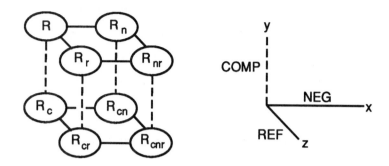

FIGURE 3.6 A rule cluster, two complementary sets of four equivalent rules.

3.3.8 Rule Clusters

Every rule R has a distinct complement R_c. The 4-rule equivalence class relates to a complementary 4-rule equivalence class, resulting in an 8-rule cluster. We depict the cluster as a box with the rules at each corner (Fig. 3.6), with complementary links along the y axis (dashed line), negative links along the x axis, and reflection links along the z axis.

The rules in a rule cluster are described by two basin fields, one for the top and one for the bottom layer. The outcome of transformations may produce the same rule, resulting in more than one occurrence of that rule in the rule cluster. In this case the cluster is shown collapsed, so that each rule only occurs once. For the purpose of reference, the lowest rule number identifies the cluster, and is positioned in the top left-hand corner. The rule clusters for all $n = 3$ (elementary rules) are set out below.

SYMMETRIC RULE CLUSTERS ($T_6 = T_3$ and $T_4 = T_1$). By definition, $R = R_r$, so the reflection links (z axis) will collapse.

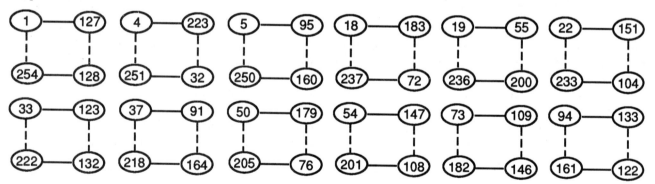

The cluster will collapse further, if for a given rule R, $R_c = R_n$,

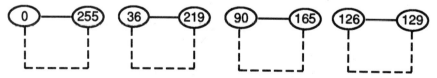

and also if $R = R_n$,

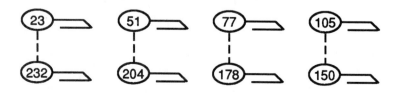

3.3 Rule Clusters

SEMI-ASYMMETRIC RULE CLUSTERS (either $T_6 \neq T_3$ or $T_4 \neq T_1$). There are no collapsed clusters among the semi-asymmetric rules.

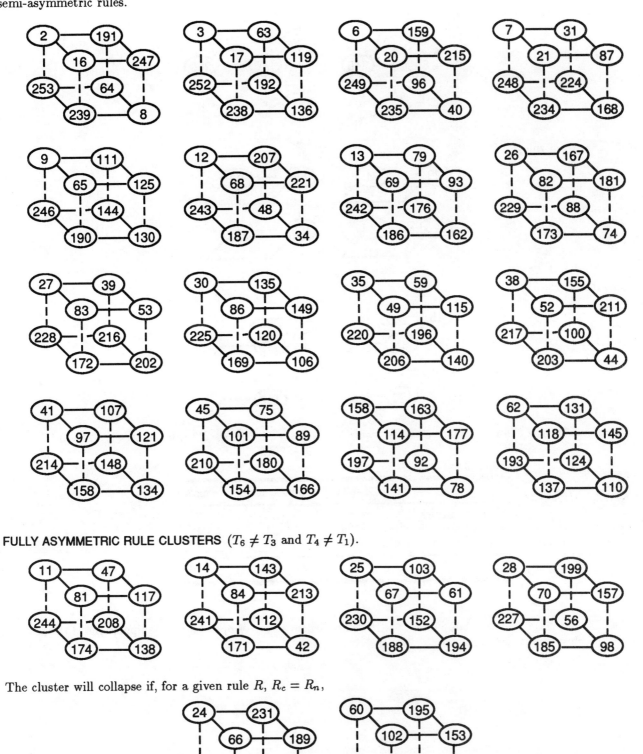

FULLY ASYMMETRIC RULE CLUSTERS ($T_6 \neq T_3$ and $T_4 \neq T_1$).

The cluster will collapse if, for a given rule R, $R_c = R_n$,

(continued)

24 THREE The Transition Function and Global Dynamics

Fully asymmetric rule clusters (continued):

And if, for a given rule R, $R_c = R_{nr}$,

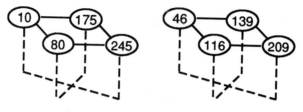

And if, for a given rule R, $R_n = R_r$,

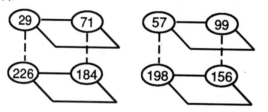

And also if $R = R_n$,

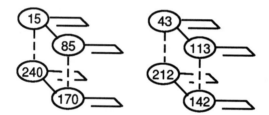

To summarise, the 256 rules collapse into 88 equivalence classes[20,28,42] and 48 rule clusters in Table 3.1.

TABLE 3.1

	rules	equiv. classes	clusters
Symmetric rules	64	36	20
Semi-asymmetric rules	128	32	16
Fully asymmetric rules	64	20	12
Total	256	88	48

The 256 rules have been tabulated on a 16 × 16 "rule-space matrix" (Appendix 4). Manipulations of the matrix simulate the clustering transformations.

3.3 Rule Clusters

3.3.9 Rule Numbering System, n=5

The $k = 2$, $n = 5$ rules have the form,

$$P_i^{t+1} = f\left(P_{i-2}^t, P_{i-1}^t, P_i^t, P_{i+1}^t, P_{i+2}^t\right)$$

where $P_i = 0$ or 1, i is the spatial position between 1 and L, and t is the time. For a circular array, length L, $P_1 = P_{L+1}, P_2 = P_{L+2}, P_3 = P_{L+3}$.

The relationship between the target cell T and the neighbourhood, including periodic boundary conditions, is shown below.

```
t0   -  -  a1 a2 a3 a4 a5 -  -         the neighbourhood a1...a5 at t0
t1   -  -  -  -  T  -  -  -  -         defines the cell T at t1

t0   a2 a3 a4 a5 -  -  -  a1     and   a5 -  -  -  a1 a2 a3 a4
t1   -  T  -  -  -  -  -  -            -  -  -  -  -  -  T  -

t0   a3 a4 a5 -  -  -  a1 a2     and   a4 a5 -  -  -  a1 a2 a3
t1   T  -  -  -  -  -  -  -            -  -  -  -  -  -  -  T
```

The $n = 5$ *rule table* with $2^5 = 32$ entries specifies a unique rule out of a total of $2^{2^5} = 2^{32} = 4{,}294{,}967{,}296$ rules. The rule table is ordered in descending values of the binary neighbourhood strings.

```
a1 -     1 1 1 1 1 1 1 1 1 1 1 1 1 1 1 1    0 0 0 0 0 0 0 0 0 0 0 0 0 0 0 0
a2 -     1 1 1 1 1 1 1 1 0 0 0 0 0 0 0 0    1 1 1 1 1 1 1 1 0 0 0 0 0 0 0 0
a3 -     1 1 1 1 0 0 0 0 1 1 1 1 0 0 0 0    1 1 1 1 0 0 0 0 1 1 1 1 0 0 0 0
a4 -     1 1 0 0 1 1 0 0 1 1 0 0 1 1 0 0    1 1 0 0 1 1 0 0 1 1 0 0 1 1 0 0
a5 -     1 0 1 0 1 0 1 0 1 0 1 0 1 0 1 0    1 0 1 0 1 0 1 0 1 0 1 0 1 0 1 0
         |                                                                |
T..      31...                                                    ...3 2 1 0
```

The rule number, R, is the decimal equivalent of the binary string $T_{31}\ldots T_0$. R will range from 0 to $2^{32} - 1$.

Note that $n = 3$ rules may be expressed as $n = 5$ rules by assigning each $n = 3$ rule table entry to four separate $n = 5$ rule table positions as specified below. For example, the $n = 3$ rule table entry T_7 is assigned to positions 31, 30, 15, 14 in the $n = 5$ rule table.

```
n = 3, T...      T7                T6                T5         ...   T0
                 |                 |                 |                |
n = 5, T...   31 = 30 = 15 = 14,  29 = 28 = 13 = 12, 27 = 26 = 11 = 10, ... 17 = 16 = 1 = 0
```

$n = 5$ rules with this rule table structure belong to the subset of the 256 $n = 3$ rules. For example, the $n = 3$ rule 193, 1100-0001, may be expressed as the $n = 5$ rule 4026789891, 1111000000000011-1111000000000011.

COMPLEMENTARY TRANSFORMATION, n=5. The $n = 5$ rules can be arranged in complementary pairs, shown linked by a vertical line,

```
if R  =     0,        1,        2,     ... (2^32/2) - 1
            |         |         |               |
Rc    =   2^32 - 1, 2^32 - 2, 2^32 - 3, ...   2^32/2
```

Two symmetries relate pairs of neighbourhoods, complementary neighbourhood pairs and reflected neighbourhood pairs, following the same reasoning as in $n = 3$ rules.

NEGATIVE TRANSFORMATION, n=5. To obtain R_n for a given rule R, in either order,
1. Take the complement of the rule, R_c.
2. Swap the output of complementary neighbourhoods pairs, shown linked by a vertical line.

$$\begin{array}{cccccc}
\overline{T}.. & 0 & 1 & 2 & 3 & \ldots & 15 \\
 & | & | & | & | & & | \\
\overline{T}.. & 31 & 30 & 29 & 28 & \ldots & 16
\end{array}$$

REFLECTION TRANSFORMATION, n=5. To obtain R_r for a given rule R, swap the outputs of the pairs of asymmetric reflected neighbourhoods, shown linked by a vertical line (note that 3/4 of neighbourhoods are asymmetric):

$$\begin{array}{ccccccccccccc}
T.. & 30 & 29 & 28 & 26 & 25 & 24 & 22 & 20 & 18 & 16 & 12 & 8 \\
 & | & | & | & | & | & | & | & | & | & | & | & | \\
T.. & 15 & 23 & 7 & 11 & 19 & 3 & 13 & 5 & 9 & 1 & 6 & 2
\end{array}$$

The reflection transformation allows rules to be placed into one of three symmetry catagories, as with $n = 3$ rules:

1. *Symmetric rules*: all pairs are equal; evolution will conserve bilateral symmetry from a symmetrical input line.
2. *Semi-asymmetric rules*: some, but not all, pairs are unequal.
3. *fully Asymmetric rules*: all pairs are unequal.

The rules will form rule clusters, equivalence classes and collapsed clusters, as with the three input rules.

All 1-D binary local CA rules cluster in this way. Two- and higher-dimensional CA will have more equivalents by reflection, and CA with a greater range of k will have k! equivalents by interchanging cell values, analogous to negative space-time patterns in binary CA.

3.3.10 Rule Numbering System, n=5 Totalistic Rules (Totalistic Code)

Totalistic rules are a small subset of the $n = 5$ rules where the value of the target cell depends only on the *total* of 1s in the neighbourhood. In $n = 5$ totalistic rules, the total of 1s can range from 0 to 5. These rules were investigated by Wolfram.[34] Following his *totalistic code* convention, the code table, with six entries, is made up of neighbourhoods that are arranged in descending order of the arithmetical sum of 1s. The neighbourhoods are shown together with their decimal equivalents.

			28 – 11100	3 – 00011		
			26 – 11010	5 – 00101		
			25 – 11001	6 – 00110		
			22 – 10110	9 – 01001		
			21 – 10101	10 – 01010		
		30 – 11110	19 – 10011	12 – 01100	1 – 00001	
		29 – 11101	14 – 01110	17 – 10001	2 – 00010	
		27 – 11011	13 – 01101	18 – 10010	4 – 00100	
		23 – 10111	11 – 01011	20 – 10100	8 – 01000	
neighbourhoods	31 – 11111	15 – 01111	7 – 00111	24 – 11000	16 – 10000	0 – 00000
total of 1s	5	4	3	2	1	0
code table	T_5	T_4	T_3	T_2	T_1	T_0

There are $2^6 = 64$ totalistic rules. The binary string $T_5 \ldots T_0$, with value 000000 to 111111, and decimal equivalents 0 to 63, is the totalistic code C. To transform a totalistic code table to an $n = 5$ rule table, the code table outputs are assigned to neighbourhoods as listed above.

Given an $n = 5$ rule table, if $T \ldots (15 = 23 = 27 = 29 = 30)$ and $(7 = 11 = 13 = 14 = 19 = 21 = 22 = 25 = 26 = 28)$ and $(3 = 5 = 6 = 9 = 10 = 12 = 17 = 18 = 20 = 24)$ and $(1 = 2 = 4 = 8 = 16)$, then the rule is totalistic.

3.4 Limited Pre-image Rules

All totalistic rules are *symmetric rules* ($C = C_r$), because the total of 1s in pairs of reflected neighbourhoods must be equal. Totalistic rules thus conserve bilateral symmetry in their space-time patterns. Complementary and negative transformations are contained within the set of totalistic rules.

NEGATIVE TRANSFORMATION, n=5 TOTALISTIC CODE. To obtain C_n for a given code C, in either order,

1. Take the complement of the code, C_c.
2. Swap the output of complementary *totals* (totals of 0s and totals of 1s).

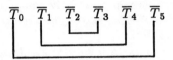

For example, code 53-110101 is transformed to code 20-010100.

TOTALISTIC CODE CLUSTERS. The 64 totalistic codes can be organised into 36 *equivalence classes* and 20 *symmetric code clusters* as shown below. Codes on the top or bottom layer of a cluster are equivalent, and will produce negative space-time patterns from a negative input line; the topology of their basin fields will be identical. Codes 0 and 63 are the same rules as the $n = 3$ rules 0 and 255.

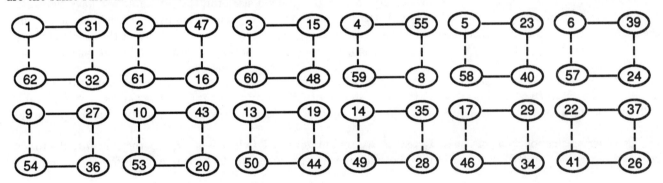

The cluster collapses further where, for a given rule C, $C_c = C_n$

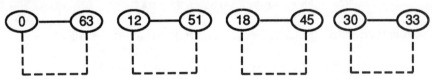

and also if $C = C_n$

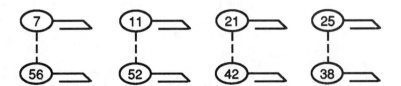

3.4 Limited Pre-image Rules

The CA rule determines the one successor state to a given state; thus, the CA's subsequent evolution is totally determined, though supposed in general to be computationally irreducible.[35–37,40] Constructing the basin of attraction field poses the problem of finding the set of all possible global states that could have preceded a given global state, its *pre-images*[12] (or predecessors[22,42]).

This problem is recognised as being very difficult,[35,36,40] requiring, in general, the exhaustive testing of the entire state space. Such a procedure becomes impractical in terms of computer time as the array length increases beyond modest values. It has been suggested that for particular rules with "simple, algebraic, structure, an efficient inversion procedure...may exist."[38] Algebraic analysis of this possibility has been pursued by Martin et al.[22] and Jen.[9,12] Jen has shown that rule specific formulae can be obtained for the exact number of pre-images for any sequence on an infinite array.[12]

The *reverse algorithm*,[42] described in this paper, directly computes all of the pre-images of a global state or directly computes that the state is a garden-of-Eden state. The average computational performance is many orders of magnitude faster than exhaustive testing.

The simplest form of the procedure applies to a subset of rules with simple algebraic structure, the *limited pre-image rules* (*limited branching rules*).[42] They include the *additive rules*[22,33] and correspond to the subset of rules that exhibit *deterministic structure* identified among the 3-neighbour rules by Jen,[8] using a similar approach.

3.4.1 Limited Pre-image Rules in General

A rule table is a list, in a conventional order, of all possible neighbourhoods and their outputs to the target cell (in the following argument, the order is irrelevant). Given a neighbourhood size of n, the rule table will have 2^n entries (examples in section 3.6). In 1-D local binary CA, the rule table may be organised into pairs of neighbourhoods that are identical except for their *rightmost* value; this identical segment is the *left start string* of the neighbourhood (conversely, the *right start string*). Such a pair of neighbourhoods may have the same or different outputs to the target cell.

If all the pairs of such neighbouhoods in the rule table are of the type that have different outputs, then the rule is defined as a *limited pre-image rule*. Note that in a limited pre-image rule the number of 0s and 1s in the rule table will, by definition, be equal, but a rule table with equal numbers of 0s and 1s is not necessarily a limited pre-image rule.

Consider a neighbourhood of size n, $a_1, a_2, \ldots, a_n$, and target cell T (where $\longrightarrow$ signifies the output to the target cell, and $\overline{T}$ signifies **not** T). If a pair of neighbourhoods with the same *left start string* have *different* outputs...

$$a_1, a_2, \ldots, a_{n-1}, 1 \longrightarrow T$$
$$\text{and} \quad a_1, a_2, \ldots, a_{n-1}, 0 \longrightarrow \overline{T},$$

then each of the pair of rule table entries has a *left deterministic permutation*, because

$$a_n \text{ is determined, given } a_1, a_2, \ldots, a_{n-1} \text{ and } T.$$

Conversely, if a pair of neighbourhoods with the same *right start string* have different outputs

$$1, a_2, a_3, \ldots, a_n \longrightarrow T$$
$$\text{and} \quad 0, a_2, a_3, \ldots, a_n \longrightarrow \overline{T},$$

then each of the pair of rule table entries has a *right deterministic permutation*, because

$$a_1 \text{ is determined, given } a_2, a_3, \ldots, a_n \text{ and } T.$$

Following Jen's terminology[8] a rule table where *all* entries have a deterministic permutation (either left, right, or both), has *deterministic structure* of the corresponding direction. A rule with, say, left deterministic structure allows a simple inverse procedure for deriving pre-images.

If the *left start string* of the candidate pre-image line, $a_1, a_2, \ldots, a_{n-1}$, is assumed, its continued derivation from *left to right* is completely determined, and its validity decided only by whether or not it complies with periodic boundary conditions. (The converse procedure is from *right to left*).

3.4 Limited Pre-image Rules

The length of the start string equals $n-1$. The number of possible start strings that need to be assumed is therefore 2^{n-1}, so this number of iterations of the procedure is required, resulting in, at most, one pre-image per iteration, irrespective of the length L of the array.

For limited pre-image rules, the maximum number of pre-images to any state in the basin field (*maximum pre-imaging*) is therefore 2^{n-1}, irrespective of L. However, it is usually less. To exhibit the maximum number of pre-images, a rule must conform to both the left *and* right deterministic structure. Rules with such *two-way* deterministic structure overlap with the much investigated[14,22,33] $n=3$ *additive* rules,[2] that have proved amenable to algebraic analysis.[22] Such rules have highly regular and predictable unfolding of their basin fields with increasing array length L, and basin field topology is highly sensitive to the number theoretic properties of L.

The vast majority of $n>3$ limited pre-image rules have either left or right deterministic structure, but not both. Such rules with *one-way* deterministic structure, say *right*, must necessarily have a rule table where some permutations of the *left start string* and T do not appear. Such *excluded permutations* will occur in conjunction with two *ambiguous permutations*, where a pair of neighbourhoods with the same *left start string* have the *same* output; a different output combined with the start string must be excluded. For example,

$$\text{if} \quad a_1, a_2, \ldots, a_{n-1}, 0 \longrightarrow T$$
$$\text{and} \quad a_1, a_2, \ldots, a_{n-1}, 1 \longrightarrow T \quad \text{(the same output)},$$

then each of the pair of rule table entries is described as having a *left ambiguous permutation*, because,

given $a_1, a_2, \ldots, a_{n-1}$ and T, a_n could be 0 or 1 with equal validity;

consequently, the permutation

$$a_1, a_2, \ldots, a_{n-1} \longrightarrow \overline{T}$$

does not appear in the rule table, and is referred to as a *left excluded permutation*.

Conversely, if a pair of neighbourhoods with the same *right start string* have the same output, that is

$$\text{if} \quad 0, a_2, a_3, \ldots, a_n \longrightarrow T$$
$$\text{and} \quad 1, a_2, a_3, \ldots, a_n \longrightarrow T \quad \text{(the same output)},$$

then each of the pair of rule table entries is described as having a *right ambiguous permutation*, because,

given $a_2, a_3, \ldots, a_n$ and T, a_1 could be 0 or 1 with equal validity;

consequently, the permutation

$$a_2, a_3, \ldots, a_n \longrightarrow \overline{T}$$

does not appear in the rule table, and is referred to as a *right excluded permutation*.

In a *one-way* (say *left*) limited pre-image rule (with *right* ambiguous permutations), maximum pre-imaging must be *less* than 2^{n-1}, irrespective of array size. A limited pre-image rule, by definition, has an equal number of 1s and 0s in its rule table; therefore, for every pair of ambiguous permutations with an output of 1, there must be another pair with an ouput of 0. For each pair of ambiguous permutations, there is a corresponding excluded permutation. Therefore, a one-way limited pre-image rule must have at least two excluded permutations with different outputs, 0 and 1, in its rule table.

To derive pre-images of a global CA state (*left to right*), the 2^{n-1} start strings $a_1, a_2, \ldots, a_{n-1}$ are assumed in turn. However, as has been shown, there will be at least two *right excluded permutations* of the form

$$a_2, a_3, \ldots, a_n \longrightarrow T$$
$$\text{and} \quad a_2, a_3, \ldots, a_n \longrightarrow \overline{T}.$$

Whatever the value of T, at least one start string is therefore invalid, and maximum pre-imaging for a *one-way* limited pre-image rule must be less than 2^{n-1}.

---

[2] An example of an $n=3$ additive rule is rule 90, $T = a_1 + a_3 \bmod 2$. Other rules considered additive are 150 and 204. Rule 60 is also additive because $T = a_1 + a_2 \bmod 2$.

In general, CA rules that are not limited pre-image rules will contain a mixture of deterministic and ambiguous (thus also excluded) permutations (both left and right), and maximum pre-imaging will increase, often exponentially, with increasing array size. Ambiguous permutations will amplify the rate of increase; deterministic and excluded permutations will inhibit the rate of increase.

The general reverse algorithm [42] (see section 3.5) applicable to local binary CA architecture uses a combination of these mechanisms, together with the idea of *partial pre-images* (unfinished, valid so far, start segments), also called partial predecessors,[42] to compute all pre-images of a given CA global state for any size of array, provided that the number of partial pre-images is within the limits of the computer system's memory.

3.4.2 Limited Pre-image Rules, n=3

In a 3-neighbour rule table, neighbourhoods that are identical except for their rightmost value, with the same *left start string*, are paired as follows:

$$\begin{array}{c|cccccccc} & 111 & 110 & 101 & 100 & 011 & 010 & 001 & 000 & \text{neighbourhoods} \\ \text{Rule table..} & T_7 & T_6 & T_5 & T_4 & T_3 & T_2 & T_1 & T_0 & \text{outputs} \end{array}$$

Part of the space-time pattern for one iteration of the CA is represented as

$$\begin{array}{llllllll} t_0 & - & - & \mathbf{A} & \mathbf{B} & \mathbf{C} & - & \text{The neighbourhood } \mathbf{ABC} \text{ at } t_0 \\ t_1 & - & - & - & \mathbf{T} & - & - & \text{defines the cell } \mathbf{T} \text{ at } t_1. \end{array}$$

If, say, $T_7 \neq T_6$, and the *left start string* $\mathbf{A\,B} = 1\,1$, $\mathbf{A\,B}$ and $\mathbf{T}$ determine $\mathbf{C}$. For instance,

if $T_7 = 0, T_6 = 1$, and $\mathbf{T} = 0$, then $\mathbf{C} = 1$;
if $T_7 = 0, T_6 = 1$, and $\mathbf{T} = 1$, then $\mathbf{C} = 0$;
if $T_7 = 1, T_6 = 0$, and $\mathbf{T} = 0$, then $\mathbf{C} = 0$;
if $T_7 = 1, T_6 = 0$, and $\mathbf{T} = 1$, then $\mathbf{C} = 1$;

This covers all possibilities, so each of the pair of rule table entries, where the left start string equals 1 1, is a *left deterministic permutation*.

If all pairs are of the type that have different outputs, i.e.,

$$T_7 \neq T_6 \text{ and } T_5 \neq T_4 \text{ and } T_3 \neq T_2 \text{ and } T_1 \neq T_0,$$

then all rule table entries are *left deterministic permutations*. In these circumstances, any combination of values of $\mathbf{A\,B}$ and $\mathbf{T}$ determine $\mathbf{C}$; the rule has *left deterministic structure* and is a limited pre-image rule. This permits pre-images of any given state to be easily derived.

To derive a pre-image of a known global state, a two-cell start segment of the pre-image line (the *partial pre-image*) is assumed, say 0 0. The rule table is consulted to find the known value of $\mathbf{T}$ in combination with $\mathbf{A\,B} = 0\,0$, giving a unique value of $\mathbf{C}$, say 1, the value of the next cell in the partial pre-image. The partial pre-image has been extended to 0 0 1. The procedure is repeated for $\mathbf{A\,B} = 0\,1$ to determine $\mathbf{C}$, the value of the next cell, and so on until the whole line plus two extra cells are derived.

If the two extra cells equal the assumed start segment, 0 0, then the pre-image is valid, because periodic boundary conditions are satisfied. Otherwise, the pre-image is not valid. As there are four possible two-cell start segments (00, 01, 10, 11), there are a maximum of four pre-images (2^{3-1}) in an $n = 3$ limited pre-image rule, irrespective of the array size.

Conversely, pre-images may be derived in the opposite direction, from right to left. The rule table, with pairs of neighbourhoods that are identical except for their *leftmost* value, with the same *right start string*, are paired as follows:

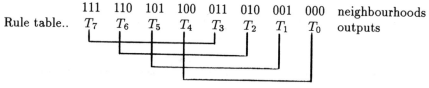

3.4 Limited Pre-image Rules

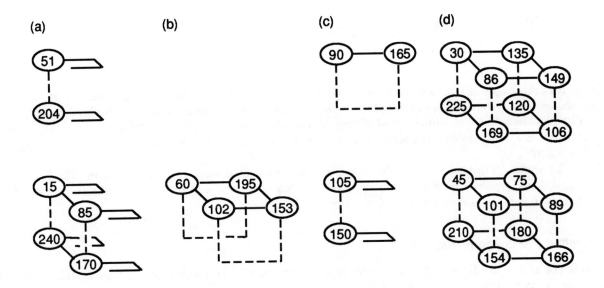

FIGURE 3.7 The set of limited pre-image rule clusters, $n \leq 3$. (a) $n = 1$, two-way, $mp = 1$; (b) $n = 2$, two-way, $mp = 2$; (c) $n = 3$, two-way, $mp = 4$; (d) $n = 3$, one-way, $mp = 3$.

An equivalent argument can be made for *right* (to *left*) deterministic permutations, deterministic structures, and limited pre-image rules.

An example of a rule with *two-way* deterministic structure, both left and right, is rule 150. Its rule table is shown below. Pairs of neighbourhoods with the same *left start string* are indicated with a dotted line (*right start string*, a solid line).

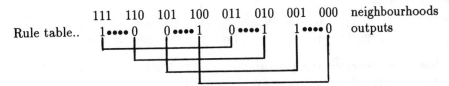

Rule 150 has exactly four pre-images for all states (other than garden-of-Eden states), for all array lengths exactly divisible by three (the neighbourhood size). Basin fields for all other array lengths consist only of attractor cycles without transients.[3]

An example of a rule with *one-way*, in this case *right*, deterministic structure is rule 30.

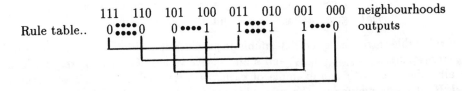

There are two pairs of *left* ambiguous permutations (indicated with a double dotted line), which must have different outputs, as the total of 0s and 1s in the rule table is equal (by definition). Thus, there are two *left excluded*

---

[3]Can a state cycle without transients justify the name *attractor*? After all, it does not *attract* anything outside of itself. We have chosen not to address this problem in terminology.

permutations with different outputs, (if "*" is a wildcard, equal to 0 or 1 with equal validity),

the ambiguous permutations are.. $11* \longrightarrow 0$ and $01* \longrightarrow 1$
and the excluded permutations are... $11* \longrightarrow 1$ and $01* \longrightarrow 0$

To derive pre-images of a global CA state (*right* to *left*), the four possible start segments, 00, 01, 10, and 11, are assumed in turn. Whatever the global state, however, one of these assumptions must be invalid, so maximum pre-imaging is 3. In rule 30, maximum pre-imaging is generally 2; however, for the global states 11111... (all 1s), pre-imaging is 3 for array sizes on the series 3, 6, 12, etc. For further examples of one-way limited pre-image rules, see section 3.5.3.

Figure 3.7 shows all limited pre-image rules among the 3-neighbour rules in their respective *rule clusters*[42] (see section 3.3) and specifies *maximum pre-imaging, mp*. Note that $n = 1$ and $n = 2$ rules are a subsets of, and expressed as, $n = 3$ rules. $n = z$ are always a subset of $n > z$ rules.

3.4.3 Limited Pre-image Rules, n=5

An example of an $n = 5$ rule with *two-way* deterministic structure is the totalistic rule, code 21. The rule table is shown below, in two parts, one below the other.

```
T.. 31 − 16   01 10  10 01  10 01  01 10
T.. 15 − 0    10 01  01 10  01 10  10 01
```

If horizontally adjacent pairs (as grouped above) are unequal, these neighbourhoods have a *left* deterministic permutation; if vertically adjacent pairs are unequal, the neighbourhoods have a *right* deterministic permutation.

Code 21 has exactly 16 pre-images (the maximum) for all states (other than garden-of-Eden states), for all array lengths exactly divisible by five (the neighbourhood size). Basin fields for all other array lengths consist only of attractor cycles without transients.

An example of an $n = 5$ rule with *one-way* (left) deterministic structure is rule 1771465110. Its rule table is shown below in the same format.

```
T.. 31 − 16   01 10  10 01  10 01  01 10
T.. 15 − 0    01 10  10 01  10 01  01 10
```

The rule table has 32 left deterministic permutations and 32 right ambiguous permutations, and is based on code 21 but with complementry rule table entries for $T_{15} - T_0$.

3.4.4 Deterministic Permutations, K>2

Deterministic permutations may be generalised for any value range k. The permutation is deterministic (left or right) if, for a neighbourhood size n,

left given $a_1, a_2, \ldots, a_{n-1}$ and T, a_n is determined;
right given $a_2, a_3, \ldots, a_n$ and T, a_1 is determined.

To determine if rule table entries have *left* deterministic permutation, the outputs of the set of k neighbourhoods with the same left start string, $a_1, a_2, \ldots, a_{n-1}$, and k alternative values of a_n (*the k-set*) are examined. If the k-set has outputs that all differ, i.e., with one example of each permitted value in the k-set, then it is termed a *left deterministic k-set*. If all k-sets making up the rule table are left deterministic, then the rule is a *left* limited pre-image rule, with maximum pre-imaging less than k^{n-1} irrespective of array size (the converse argument applies to a *right deterministic k-set* and *right* limited pre-image rules). If the rule has two-way deterministic structure (left *and* right), then pre-imaging may equal the maximum of k^{n-1}.

For instance, if $n = 3$ and $k = 3$ (with values 0, 1 and 2), then a segment of the rule table, consisting of a left deterministic k-set, may appear as follows

```
002  001  000   neighbourhoods
 0    1    2    outputs
```

where each permutation is deterministic, whereas the non-deterministic k-set

$$\begin{array}{ccc} 002 & 001 & 000 \\ 1 & 1 & 2 \end{array} \quad \begin{array}{l} \text{neighbourhoods} \\ \text{outputs} \end{array}$$

consists of two ambiguous permutations $002 \to 1$ and $001 \to 1$ and one deterministic permutation $000 \to 2$.

The example above demonstrates that if $k > 2$, deterministic permutations may exist within a non-deterministic k-set (This is not the case with a $k = 2$, binary CA). The number of deterministic permutations in a k-set depends on the distribution of the k values to the k outputs. If some values are missing, resulting in the same values assigned to several outputs, then there must be some ambiguous (and some excluded) permutations. The *degree of ambiguity* of ambiguous permutations depends on how many neighbourhoods in the k-set share the same output. If only part of the k values are used for the outputs to a k-set, then the number of deterministic permutations in the k-set will equal the number of outputs that occur once only.

As with $k = 2$ rules, ambiguous permutations will amplify the rate of increase of pre-imaging with increasing array size (with greater amplification for a greater degree of ambiguity); deterministic and excluded permutations will inhibit the rate of increase.

3.5 The Reverse Algorithm

This section describes the logic for directly generating pre-images, starting with the *limited pre-image rules*. It was shown in section 3.4 that limited pre-image rules have a fixed maximum number of pre-images, mp, to any node irrespective of array length. For a neighbourhood size n, $mp \leq 2^{n-1}$. The rules may be expressed as simple algorithms relating the target cell to neighbourhood values; these rule algorithms are included below. The limited pre-image rules are examined in the order of increasing n.

3.5.1 Pre-images of n=1 Limited Pre-image Rules

$$\begin{array}{ll} \text{Consider the neighbourhood...} & \textbf{A B C} \\ \text{and the target cell} & \textbf{T} \end{array}$$

The trivial $n = 1$ rule subset, expressed as $n = 3$ rules, has neighbourhoods as follows (where "$*$" signifies a wild card, a member of the $n = 3$ neighbourhood that is irrelevant to the target)

$$* \textbf{B} *, \quad \textbf{A} * *, \quad * * \textbf{C}.$$

The $n = 1$ rule table consists of 2 entries.

$$\text{Rule table..} \quad \begin{array}{cc} 1 & 0 \\ T_1 & T_0 \end{array} \quad \begin{array}{l} \text{neighbourhoods} \\ \text{outputs} \end{array}$$

If $T_1 = T_0$ (an ambiguous permutation), then the rule belongs to the trivial $n = 0$ rule cluster (a subset of $n = 1$). The rule cluster 0 is shown below, where every state in state space is the pre-image of one of the two *uniform states*, all 0s or 1s.

SYMMETRICAL CLUSTER 0.

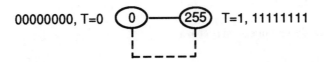

00000000, T=0   0 —— 255   T=1, 11111111

If $T_1 \neq T_0$, the rule has deterministic structure.

THREE The Transition Function and Global Dynamics

Consider the $n = 3$ rule table:

	111	110	101	100	011	010	001	000	neighbourhoods
Rule table..	T_7	T_6	T_5	T_4	T_3	T_2	T_1	T_0	outputs

Matching wildcard neighbourhoods with the $n = 3$ rule table, we obtain the following $n = 3$ *deterministic structures* for $n = 1$ rules.

$\quad$ * **B** * $\quad$ rules....$(T_7 = T_6 = T_3 = T_2) \neq (T_5 = T_4 = T_1 = T_0)$;
$\quad$ **A** * * $\quad$ rules....$(T_7 = T_6 = T_5 = T_4) \neq (T_3 = T_2 = T_1 = T_0)$;
$\quad$ * * **C** $\quad$ rules....$(T_7 = T_5 = T_3 = T_1) \neq (T_6 = T_4 = T_2 = T_0)$.

Rules that satisfy these conditions are $n = 1$ limited pre-image rules. They are shown below expressed as $n = 3$ rules, in their respective rule clusters, with the rule table, rule number, and rule algorithm.

SYMMETRICAL CLUSTER 51.

(B), 00110011, T≠B $\quad$ 51

(B), 11001100, T=B $\quad$ 204

FULLY ASYMMETRIC CLUSTER 15.

(A), 00001111, T≠A $\quad$ 15

$\qquad\qquad\qquad\qquad$ 85 $\quad$ T≠C, 01010101, (C)

(A), 11110000, T=A $\quad$ 240

$\qquad\qquad\qquad\qquad$ 170 $\quad$ T=C, 10101010, (C)

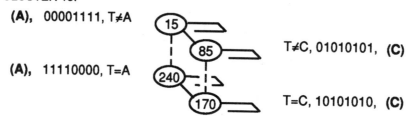

The behaviour and basin field structure of $n = 1$ rules is predictable and $n = 1$ limited pre-image rules are reversible. Every state must have one pre-image; therefore, there are no garden-of-Eden states nor transients, and all states belong to attractor cycles.

3.5.2 Pre-images of n=2 Limited Pre-image Rules

$\qquad$ Consider the neighbourhood... $\quad$ **A B C**
$\qquad$ and the target cell $\qquad\qquad\qquad$ **T**

Expressed as $n = 3$ rules, the $n = 2$ rules have neighbourhood as follows, where "*" signifies a wild card:

$$\mathbf{A\,B\,*}, \quad \mathbf{*\,B\,C}.$$

The $n = 2$ rule table consists of four entries.

	1 1	1 0	0 1	0 0	neighbourhoods
Rule table..	T_3	T_2	T_1	T_0	outputs

Given this rule table, a rule has deterministic structure

$$\text{if } (left) \; T_3 \neq T_2 \text{ and } T_1 \neq T_0,$$
$$\text{or } (right) \; T_3 \neq T_1 \text{ and } T_2 \neq T_0.$$

3.5 The Reverse Algorithm

Consider the $n=3$ rule table:

$$\begin{array}{c|cccccccc}
 & 111 & 110 & 101 & 100 & 011 & 010 & 001 & 000 \quad \text{neighbourhoods} \\
\text{Rule table..} & T_7 & T_6 & T_5 & T_4 & T_3 & T_2 & T_1 & T_0 \quad \text{outputs}
\end{array}$$

Matching wildcard neighbourhoods with the $n=3$ rule table, we obtain the following $n=3$ deterministic structures for $n=2$ rules.

left   **A B** *   rules.......$(T_7 = T_6) \neq (T_5 = T_4)$ and $(T_3 = T_2) \neq (T_1 = T_0)$;
right   * **B C**   rules.......$(T_7 = T_3) \neq (T_5 = T_1)$ and $(T_6 = T_2) \neq (T_4 = T_0)$.

Looking back at the $n=1$ rules, cluster 51 satisfies both conditions, and cluster 15 satisfies one or the other condition.

The $n=2$ rules that satisfy these conditions are limited pre-image rules. They are shown below expressed as $n=3$ rules, in their respective rule clusters, with the rule table, rule number, and rule algorithm.

DOUBLY ASYMMETRIC CLUSTER 60.

(AB),   00111100, if A=B, T=0 else 1   if A=B, T=1 else 0, 11000011

(BC),   01100110, if B=C, T=0 else 1   if B=C, T=1 else 0, 10011001

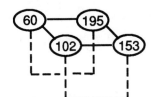

Note that the $n=2$ rule table for these rules has *two-way* deterministic structure, but the $n=3$ rule table has *one-way* deterministic structure. For **A B** rules, **A** and **T** will determine **B** (left to right); conversely, for **B C** rules, **B** and **T** will determine **C** (right to left).

Consider an unknown pre-image, $P_0, P_1, P_2, \ldots, P_L, P_{L+1}$ (with notional values P_0 and P_{L+1}) of the known successor line of length L, $I_1, I_2, \ldots, I_L$. As boundary conditions are periodic, $P_0 = P_L$ and $P_{L+1} = P_1$. To derive pre-images, the value of P_0 is assumed, and the rule table used to determine, from left to right, successive values of P_x according to P_{x-1} and I_x.

$$\begin{array}{cccccccc}
P_0 & P_1 & P_2 & \ldots & P_{x-1} & P_x & \ldots & P_{L-1} & P_L \\
 & I_1 & I_2 & \ldots & & I_x & \ldots & & I_L
\end{array}$$

When the final value P_L is derived, if $P_L = P_0$, then periodic boundary conditions are satisfied, and the pre-image is valid. As P_0 can have two assumed values, 0 and 1, each is assumed in turn to generate, at most, one pre-image, resulting in a maximum of two possible pre-images.

Conversely, pre-images may be derived by assuming P_{L+1}, and determining, from *right* to *left*, successive values of P_x according to P_{x+1} and I_x.

$$\begin{array}{cccccccc}
 & P_1 & P_2 & \ldots & P_x & P_{x+1} & \ldots & P_{L-1} & P_L & P_{L+1} \\
 & I_1 & & \ldots & & I_x & \ldots & & I_{L-1} & I_L
\end{array}$$

If $P_1 = P_{L+1}$, the pre-image is valid, and the same conclusions apply as with the left to right procedure.

3.5.3 Pre-images of n=3 Limited Pre-image Rules

Consider the neighbourhood...   **A B C**
and the target cell   **T**

THREE The Transition Function and Global Dynamics

Consider the $r = 3$ rule table:

	111	110	101	100	011	010	001	000	neighbourhoods
Rule table..	T_7	T_6	T_5	T_4	T_3	T_2	T_1	T_0	outputs

As shown in section 3.4, deterministic structures for $n = 3$ rules (left, right or two-way) are as follows:

    left   (**AB** rules)..........$T_7 \neq T_6$ and $T_5 \neq T_4$ and $T_3 \neq T_2$ and $T_1 \neq T_0$;
    right   (**BC** rules)..........$T_7 \neq T_3$ and $T_5 \neq T_1$ and $T_6 \neq T_2$ and $T_4 \neq T_0$;
    two-way (**ABC** rules).......both of the above conditions are true.

Looking back at the $n < 3$ rules, the $n = 1$ cluster 15 and the $n = 2$ cluster 60 are either **AB** or **BC** rules. The $n = 3$ limited pre-image rules are set out below.

SYMMETRIC CLUSTER 90.

    (ABC),   01011010, if A=C, T=0 else 1   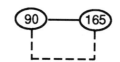   if A=C, T=1 else 0, 10100101

SYMMETRIC CLUSTER 105.

    (ABC),   01101001, if A≠C, T=B else $\overline{B}$ (105)

    (ABC),   10010110, if A=C, T=B else $\overline{B}$ (150)

SEMI-ASYMMETRIC CLUSTER 30.

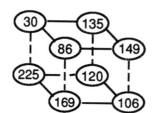

(BC),	00011110, if BC=00, T=A else $\overline{A}$		if BC=11, T=A else $\overline{A}$, 10000111
(AB),	01010110, if AB=00, T=C else $\overline{C}$		if AB=11, T=C else $\overline{C}$, 10010101
(BC),	11100001, if BC=00, T=$\overline{A}$ else A		if BC=11, T=$\overline{A}$ else A, 01111000
(AB),	10101001, if AB=00, T=C else $\overline{C}$		if AB=11, T=$\overline{C}$ else C, 01101010

SEMI-ASYMMETRIC CLUSTER 45.

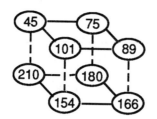

(BC),	00101101, if B<C, T=A else $\overline{A}$		if B>C, T=A else $\overline{A}$, 01001011
(AB),	01100101, if A>B, T=$\overline{C}$ else C		if A<B, T=C else $\overline{C}$, 01011001
(BC),	11010010, if B<C, T=$\overline{A}$ else A		if B>C, T=$\overline{A}$ else A, 10110100
(AB),	10011010, if A>B, T=C else $\overline{C}$		if A<B, T=$\overline{C}$ else C, 10100110

The rules have deterministic structure of the following type,

    left   **AB** rules, **AB** and **T** will determine **C**;
    right   **BC** rules, **BC** and **T** will determine **A**;
    two-way **ABC** rules, both the above will apply.

3.5 The Reverse Algorithm

To determine the set of pre-images of a given state for **A B** rules, consider an unknown pre-image, $P_0, P_1, P_2, \ldots, P_L, P_{L+1}$ (with notional values P_0 and P_{L+1}) of the known successor line of length L, $I_1, I_2, \ldots, I_L$. As boundary conditions are periodic, $P_0 = P_L$ and $P_{L+1} = P_1$. To derive pre-images, the value of the start string, $P_0 P_1$, is assumed and the rule table used to determine, from left to right, successive values of P_{x+1} according to P_{x-1}, P_x, and I_x.

$$P_0 \quad P_1 \quad P_2 \quad P_3 \quad \ldots \quad P_{x-1} \quad P_x \quad P_{x+1} \quad \ldots \quad P_{L-1} \quad P_L \quad P_{L+1}$$
$$I_1 \quad I_2 \quad \ldots \quad I_x \quad \ldots \quad I_L$$

If $P_L = P_0$ and $P_{L+1} = P_1$, then periodic boundary conditions are satisfied, and the pre-image is valid. As the start string, $P_0 P_1$, can have four values (00, 01, 10, 11), each is assumed in turn to generate at most one pre-image, resulting in a maximum of four possible pre-images.

Conversely, for **B C** rules, pre-images may be derived by assuming the start string, $P_L P_{L+1}$, and determining, from *right* to *left*, successive values of P_{x-1} according to P_{x+1}, P_x and I_x.

$$P_0 \quad P_1 \quad P_2 \quad P_3 \quad \ldots \quad P_{x-1} \quad P_x \quad P_{x+1} \quad \ldots \quad P_{L-1} \quad P_L \quad P_{L+1}$$
$$I_1 \quad I_2 \quad \ldots \quad I_x \quad \ldots \quad I_L$$

If $P_L = P_0$ and $P_{L+1} = P_1$, then the pre-image is valid, and the same conclusions apply as with the left-to-right procedure.

Clusters 30 and 45 have *one-way* deterministic structure, either left (**A B**), or right (**B C**). Such rules have *excluded permutations* that reduces maximum pre-imaging to 3 (see section 3.4).

Consider the neighbourhood... **A B C**
and the target cell **T**

and the rule tables for the **A B** rule 86 (cluster 30) and 101 (cluster 45), which have *left* deterministic structure.

Rule table..	111	110	101	100	011	010	001	000	neighbourhoods
	T_7	T_6	T_5	T_4	T_3	T_2	T_1	T_0	outputs
	0	1	0	1	0	1	1	0	rule 86
	0	1	1	0	0	1	0	1	rule 101

Every permutation of **A B** and **T** will give a unique value of **C**. However there are two permutations of **B C** and **T** which are forbidden, $\begin{smallmatrix}1&1\\&1\end{smallmatrix}$ and $\begin{smallmatrix}1&0\\&0\end{smallmatrix}$

Thus, for any value of **T**, there is one forbidden value of the string **B C**. To determine the set of pre-images, we assumed in turn the four alternative start strings, $P_0 P_1$. However, one of these start strings must be invalid; therefore, maximum pre-imaging = 3. An equivalent argument can be made for all the other rules in the clusters 30 and 45.

3.5.4 The Reverse Algorithm for Pre-images of Any n=3 Rule

Consider the neighbourhood... **A B C**
and the target cell **T**

Consider the $n = 3$ rule table:

Rule table..	111	110	101	100	011	010	001	000	neighbourhoods
	T_7	T_6	T_5	T_4	T_3	T_2	T_1	T_0	outputs

and the limited pre-image rules with one, or both, of the following deterministic structures:

left   (**AB** rules).......$T_7 \neq T_6$ and $T_5 \neq T_4$ and $T_3 \neq T_2$ and $T_1 \neq T_0$;
right (**BC** rules).......$T_7 \neq T_3$ and $T_5 \neq T_1$ and $T_6 \neq T_2$ and $T_4 \neq T_0$.

A deterministic structure has a complete set of *deterministic permutations*, with *unequal pairs* of outputs of the form $T_7 \neq T_6$; given a *left start string* **A B** = 11 and **T**, **C** is determined. If, however, there are *ambiguous permutations* with equal pairs of the form $T_7 = T_6$, and if **A B** and **T** is not an *excluded permutation*, then **C** could be 0 or 1 with equal validity.

Most rules have a mixture of ambiguous permutations (which amplify pre-imaging), and deterministic and excluded permutations (which inhibit pre-imaging).

To determine the set of pre-images of a given state for *any* rule, consider an unknown pre-image, $P_0, P_1, P_2, \ldots, P_L, P_{L+1}$ (with notional values P_0 and P_{L+1}) of the known successor line of length L, $I_1, I_2, \ldots, I_L$. As boundary conditions are periodic, $P_0 = P_L$ and $P_{L+1} = P_1$. To derive the values of successive cells of a candidate pre-image, the value of the start string, $P_0 P_1$, is assumed and the rule table used to determine, from left to right, successive values of P_{x+1} according to P_{x-1}, P_x, and I_x. As the start string accretes more cell values, we call it the *partial pre-image*.

$$P_0 \quad P_1 \quad P_2 \quad P_3 \quad \ldots \quad P_{x-1} \quad P_x \quad P_{x+1} \quad \ldots \quad P_{L-1} \quad P_L \quad P_{L+1}$$

$$I_1 \quad I_2 \quad \ldots \quad I_x \quad \ldots \quad I_L$$

At each step, the following procedure is enacted:

1. If P_{x-1}, P_x, and I_x make up an *excluded* permutation, abandon the start string or partial pre-image. Resume derivation of the next partial pre-image (step 5).

2. If P_{x-1}, P_x, and I_x make up a *deterministic* permutation, P_{x+1} has a unique value; proceed to the next cell (step 1).

3a. If P_{x-1}, P_x, and I_x make up an *ambiguous* permutation, P_{x+1} could be 0 or 1 with equal validity; assume 0.

3b. If P_x, P_{x+1}, and I_x make up an *excluded* permutation, change P_{x+1} to 1, and proceed to the next cell (step 1).

3c. If P_x, P_{x+1}, and I_x do *not* make up an *excluded* permutation, record the partial pre-image ending with $P_{x+1} = 1$, adding it to the *partial pre-image queue*. Reassume $P_{x+1} = 0$ and proceed to the next cell (step 1). At each new cell position, it may be necessary to add one partial pre-image to the queue.

4. Once the pre-image (ending with P_{L+1}) is completed, if periodic boundary conditions ($P_L = P_0$ and $P_{L+1} = P_1$) are *not* satisfied, the pre-image is abandoned; if satisfied, the pre-image is added to the list of valid pre-images.

5. Take the earliest partial pre-image, recorded in step 3, from the head of the partial pre-image queue and proceed to the next cell (step 1). More partial pre-images may be added to the end of the queue at step 3, and removed from the beginning of the queue at step 5, until none remain.

6. When the partial pre-image queue is exhausted, all possible pre-images, starting with the start string $P_0 P_1$, have been derived without duplication. The procedure is repeated in turn for the three remaining values of $P_0 P_1$.

The *reverse algorithm* will compute all possible pre-images, without duplication, to a given CA state for any rule and any array length, provided that the number of partial pre-images in the partial pre-image queue at any one time is within the limits of the computer system's memory. For an array size L, and a state with p pre-images, if the time required to output the full set of pre-images is t_p, it is estimated that, using this algorithm, t_p increases arithmetically with L and p (garden-of-Eden states are identified instantly), whereas, if the whole state space is exhaustively tested for pre-images, for any p, however small, t_p increases exponentially as 2^L.

3.5.5 Pre-images of n=5 Rules

Pre-images for $n = 5$ rules, or indeed $n > 5$ rules, may be computed with an extended reverse algorithm.

Consider the $n = 5$ neighbourhood... **a_1 a_2 a_3 a_4 a_5**
and the target cell **T**

and consider the 32 neighbourhoods and their outputs $T_{31}\ldots T_0$:

```
a1 -    1 1 1 1 1 1 1 1 1 1 1 1 1 1 1 1    0 0 0 0 0 0 0 0 0 0 0 0 0 0 0 0
a2 -    1 1 1 1 1 1 1 1 0 0 0 0 0 0 0 0    1 1 1 1 1 1 1 1 0 0 0 0 0 0 0 0
a3 -    1 1 1 1 0 0 0 0 1 1 1 1 0 0 0 0    1 1 1 1 0 0 0 0 1 1 1 1 0 0 0 0
a4 -    1 1 0 0 1 1 0 0 1 1 0 0 1 1 0 0    1 1 0 0 1 1 0 0 1 1 0 0 1 1 0 0
a5 -    1 0 1 0 1 0 1 0 1 0 1 0 1 0 1 0    1 0 1 0 1 0 1 0 1 0 1 0 1 0 1 0
        |                                  |
T..     31 ...                                             ...3 2 1 0
```

Deterministic structures for $n = 5$ rules (left, right, or two-way) are as follows:

left $T_{31} \neq T_{30}$ and $T_{29} \neq T_{28}$ and $\ldots T_3 \neq T_2$ and $T_1 \neq T_0$,
generally, for odd $x = 1$ to 31, $T_x \neq T_{x-1}$;

right $T_{31} \neq T_{15}$ and $T_{30} \neq T_{14}$ and $\ldots T_{17} \neq T_1$ and $T_{16} \neq T_0$,
generally, for $x = 0$ to 15, $T_{x+16} \neq T_x$.

Given a *left* deterministic permutation, the start string $a_1a_2a_3a_4$ and **T** will determine a_5. The start string is four cells long, and thus has $2^4 = 16$ permutations, so maximum pre-imaging will be 16, irrespective of array size. Conversely, given a *right* deterministic permutation, the start string $a_2a_3a_4a_5$ and **T** will determine a_1, and the same conclusions apply.

The vast majority of $n = 5$ limited pre-image rules will have *one-way* deterministic structure, and will therefore have *excluded permutations*, resulting in less than maximum pre-imaging. Generally for a neighbourhood of size n, maximum pre-imaging is less than or equal to 2^{n-1}.

The procedure for deriving the pre-images of any $n = 5$ rule, takes into account deterministic, forbidden and ambiguous permutations, in an extension of the algorithm presented in section 3.5.4. In principle, the reverse algorithm may be applied to rules with any value of n, and may be extended for $k > 2$.

3.6 The Z Parameter

We will define *maximum pre-imaging* as the greatest number of incoming arcs to any node in a basin of attraction field. Maximum pre-imaging as a function of array length (see the data in the Atlas) would be expected to tie into other quantifiable measures of basin fields, for example, the density of garden-of-Eden nodes, and would reflect the degree of *convergence* of state space, discussed further in chapter 4. It would be useful to identify a quantifiable parameter implicit in the rule table that reflected such behaviour.

It is possible to quantify the *probability* that the *next unknown cell* in a partial pre-image is determined (i.e., has one solution, either 0 or 1); this will reflect the maximum pre-imaging exhibited and its variation with array length L, characteristic of a given rule. This probability is refered to as the *Z parameter*.

Consider the reverse algorithm for computing a pre-image (either from left to right, or conversely from right to left) as described in section 3.8. The left-to-right computation is represented by the diagram below:

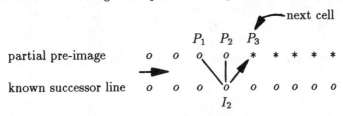

where "o" represents a known cell value, and "$*$" an unknown value.

As a first approximation, the minimum value of Z is the proportion of *deterministic permutations* in a rule table, left or right, whichever is greater. It was demonstrated in section 3.4 that, if the rule table has a *deterministic structure*, i.e., 100% deterministic permutations, then the probability that the next cell is determined equals 1, and the rule is a *limited pre-image rule*. Maximum pre-imaging, mp, is constant with increasing L.

$$Z = 1, \quad mp \leq 2^{n-1}.$$

If the rule table has no deterministic permutations, for instance, the trivial rule 0, then the probability that the next cell is determined equals 0, and mp diverges exponentially as a function of L

$$Z = 0, \quad mp = 2^L.$$

The overwhelming majority of possible rules will have a rule table that has an intermediate proportion of deterministic permutations; Z and mp will have intermediate values.

For example, the $n = 3$ rule 193 has the rule table,

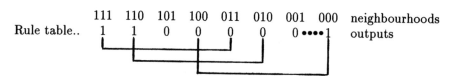

Left deterministic permutations are linked by a dotted line; the total, DP_L, equals 2. Right deterministic permutations are linked by a solid line; the total, DP_R, equals 6. The greater value as a proportion of the total of eight neighbourhoods equals 6/8, so $Z = .75$.

Maximum pre-imaging data for rule 193, taken from the Atlas, is tabulated below in relation to L, with the intermediate function $\sqrt{2^L}$ listed as a rough yardstick for assessing the divergence of mp with L.

$L..$	1	2	3	4	5	6	7	8	9	10	11	12	13	14	15
$mp..$	2	2	3	2	5	5	7	10	12	17	22	29	39	51	68
$\sqrt{2^L}$		2		4		8		16		32		64		128	

For an example of an $n = 5$ rule, consider the totalistic rule, code 50. The rule table is shown below in two parts, one below the other.

```
T.. 31 − 16   1   1   1−0   1−0   0   0   1−0   0   0   0   0   0−1
              |   |   |           |   |       |       |   |   |
T.. 15 − 0   1−0   0   0   0   0−1   0   0   0−1   0−1   1−0
```

Left deterministic permutations are linked by a horizontal dash; the total, DP_L, equals 18. Right deterministic permutations are linked vertically; the total, DP_R, equals 18. Therefore, $Z = 18/32 = 0.5625$ (note that *symmetric* rules must have the same number of left and right deterministic permutations, and all totalistic rules are symmetrical). Maximum pre-imaging data for code 50, taken from the Atlas, is tabulated below, with the intermediate function $\sqrt{2^L}$ listed as a rough yardstick for assessing the divergence of mp with L.

$L..$	3	4	5	6	7	8	9	10	11	12	13	14	15	16
$mp..$	2	7	21	21	15	31	55	123	155	205	313	549	951	1423
$\sqrt{2^L}$		4		8		16		32		64		128		256

The divergence of mp with L relates inversely to Z (see chapter 4). For instance, the divergence of mp for rule 193 ($Z = .75$) is less than $\sqrt{2^L}$, whereas the divergence of mp for code 50 ($Z = .5625$) is greater than $\sqrt{2^L}$.

3.6.1 Corrected Z Parameter: n=3

Within the rule table of, say, an $n = 3$ rule, there may be *hidden deterministic permutations* present relating to the smaller neighbourhoods, $n = 2$ and $n = 1$. Within an $n = 5$ rule table there may be hidden deterministic

3.6 The Z Parameter

permutations relating to the smaller neighbourhoods, $n = 4$, $n = 3$, $n = 2$, and $n = 1$. Generally, an $n = z$ rule table may include deterministic permutations relating to neighbourhoods smaller than z. The first approximation of the value of the Z parameter must be corrected upward in many cases to take this into account.

Consider the $n = 3$ neighbourhood... **A B C**   In a left deterministic
                                                    permutation, **AB** and **T** will
and the target cell   **T**   determine **C** (conversely, right).

The $n = 3$ rule table for rule 51, with neighbourhoods rearranged vertically, is shown below,

```
A..  1 1   1 1   0 0   0 0
B..  1 1   0 0   1 1   0 0
C..  1 0   1 0   1 0   1 0    neighbourhoods ABC

T..  0 0   1 1   0 0   1 1    rule table, rule 51
```

Although the rule table has no deterministic permutations (left or right), relating to the neighbourhood **A B C**, it has 100% *left deterministic permutations* relating to the neighbourhood **A B**, where row **C** can be ignored; thus,

the neighbourhood... **A B C**   Given **A** and **T**
                                  B is determined;
and the target cell   **T**   **C** is irrelevant.

For an $n = 3$ rule, there will be three types of deterministic permutations (left and right), labelled $n3$, $n2$, and $n1$, as tabulated below:

neighbourhood size	left	right	
$n3$	**ABC**	**ABC**	Neighbourhoods to which the
$n2$	**AB**	**BC**	deterministic permutations
$n1$	**A**	**C**	relate.

To count the number of the three types of deterministic permutations in a rule table, it is convenient to visualise a set of deterministic *templates* that are applied to the rule table to check if specific deterministic permutations exist. For $n = 3$ rules there will be a left and a right set of three templates, $n3$, $n2$, and $n1$.

n=3. The $n3$ *template* (left or right) follows the definition described in section 3.4. The template size equals 2 and occupies 1/4 of the rule table, with a maximum of four positions. (In all the following examples, left and right templates and the corresponding deterministic permutation diagrams are illustrated.)

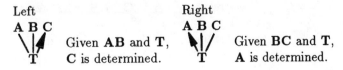

The curved link indicates rule table entries are different:

Left   Right
A B C   **A B C**
   T   Given **AB** and **T**,   **T**   Given **BC** and **T**,
           C is determined.   **A** is determined.

42 THREE The Transition Function and Global Dynamics

Let n_3 be the number of *left* $n3$ template positions. Then the probability that $P_1P_2P_3$ belongs to a neighbourhood on an $n3$ template equals $n_3/4$. If so, P_1, P_2, and I_2 will determine the *next cell* P_3, illustrated in the diagram below (conversely, *right*).

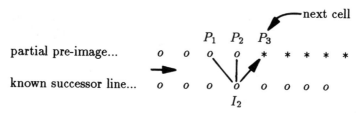

n=2. The *n2 template* (left or right), size 4, has two possible positions and occupies 1/2 the rule table.

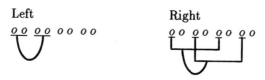

The underline (or linked underlines) indicated entries are equal; the curved link indicates the two sets of equal entries are different.

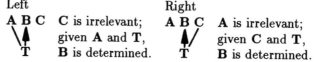

Let n_2 be the number of *left* $n2$ template positions. Then the probability that $P_2P_3P_4$ belongs to a neighbourhood on an $n2$ template equals $n_2/2$. If so, P_2 and I_3 will determine the *next cell* P_3, illustrated in the diagram below (conversely, *right*).

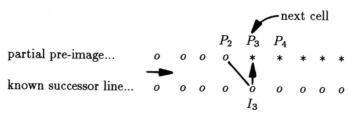

n=1. The trivial $n1$ *template* (left or right), size 8, has one possible position and occupies the entire rule table.

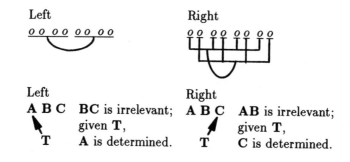

3.6 The Z Parameter

If a *left* n1 template fits, then $n_1 = 1$. The probability that $P_3P_4P_5$ belongs to a neighbourhood on the n1 template equals n_1. If so, I_4 will determine the *next cell* P_3, illustrated in the diagram below (conversely, *right*).

```
                                              ┌── next cell
                                              ↓
                                           P₃  P₄  P₅
     partial pre-image...     o  o  o  o   *   *   *   *   *
                           →
     known successor line...  o  o  o  o   o   o   o   o   o
                                              ↑
                                             I₄
```

To summarise, given an $n = 3$ rule, the number of positions where the three types of template fit the rule table are designated as follows,

$$n_3 \text{ (max 4)} \ldots \text{ n3 template (size 2)}$$
$$n_2 \text{ (max 2)} \ldots \text{ n2 template (size 4)}$$
$$n_1 \text{ (max 1)} \ldots \text{ n1 template (size 8)}$$

Templates relating to a given direction are mutually exclusive; a given rule table entry may not connect to more than one left template and one right template. The following method will give the corrected probability that the *next cell* is determined, thus the corrected value of Z. The method described below computes Z_L, from left to right. A converse, but otherwise identical method computes Z_R, from right to left.

- n3: let p_3 be the probability that the *next cell* is determined by n3, $p_3 = n_3/4$; the probability that it is *not* determined, $\overline{p}_3 = 1 - p_3$
- n2: if *not* determined by n3, let p_2 be the probability that the *next cell* is determined by n2, $p_2 = n_2/2 \times \overline{p}_3$
- n1: this is a special case, as the n1 template takes up the whole rule table. Let p_1 be the probability that the *next cell* is determined by n1, $p_1 = n_1$ (note: if $n_1 = 1$, Z_L and $Z_R = 1$)

$$Z_L = p_3 + p_2 + p_1.$$

The corrected parameter $Z = Z_L$ or Z_R, whichever is greater.

As an example, consider the $n = 3$ rule 54. The rule table is set out below; left template positions are $n_3 = 2$ and $n_2 = 1$ and right template positions are $n_3 = 2$ and $n_2 = 1$

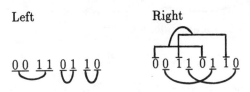

Left procedure $p_3 = n_3/4 = 2/4 = 1/2.$ $\overline{p}_3 = 1 - 1/2 = 1/2$
(same for right) $p_2 = n_2/2 \times \overline{p}_3 = 1/2 \times 1/2 = 1/4$
                  $Z_L = p_3 + p_2 = 1/2 + 1/4 = 3/4$
                  $Z_L = Z_R,$ so $Z = .75$

3.6.2 Corrected Z Parameter: n=5

Consider the rule table for $n = 5$ rules, with 32 neighbourhoods and their outputs $T_{31} \ldots T_0$:

```
a1 -   1 1 1 1 1 1 1 1 1 1 1 1 1 1 1 1    0 0 0 0 0 0 0 0 0 0 0 0 0 0 0 0
a2 -   1 1 1 1 1 1 1 1 0 0 0 0 0 0 0 0    1 1 1 1 1 1 1 1 0 0 0 0 0 0 0 0
a3 -   1 1 1 1 0 0 0 0 1 1 1 1 0 0 0 0    1 1 1 1 0 0 0 0 1 1 1 1 0 0 0 0
a4 -   1 1 0 0 1 1 0 0 1 1 0 0 1 1 0 0    1 1 0 0 1 1 0 0 1 1 0 0 1 1 0 0
a5 -   1 0 1 0 1 0 1 0 1 0 1 0 1 0 1 0    1 0 1 0 1 0 1 0 1 0 1 0 1 0 1 0
        |                                                                |
T..    31 ...                                                   ... 3 2 1 0
```

There will be five types of deterministic permutations (left and right), labelled $n5$, $n4$, $n3$, $n2$, and $n1$, as tabulated below:

neighbourhood size	left	right	
$n5$	$a_1\ a_2\ a_3\ a_4\ a_5$	$a_1\ a_2\ a_3\ a_4\ a_5$	
$n4$	$a_1\ a_2\ a_3\ a_4$	$a_2\ a_3\ a_4\ a_5$	
$n3$	$a_1\ a_2\ a_3$	$a_3\ a_4\ a_5$	Neighbourhoods to which the
$n2$	$a_1\ a_2$	$a_4\ a_5$	deterministic permutations
$n1$	a_1	a_5	relate.

n=5. The $n5$ *template* (left or right) follows the definition described in section 3.4. The template size equals 2 and occupies 1/16 of the rule table, with a maximum of 16 positions. It is shown below.

Left

o o o o o o o o   o o o o o o o o   o o o o o o o o   o o o o o o o o

Right

o o o o o o o o   o o o o o o o o   o o o o o o o o   o o o o o o o o

The curved link indicates rule table entries are different:

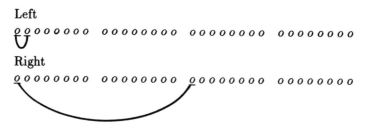

Left — Given $a_1 a_2 a_3 a_4$ and T, a_5 is determined.

Right — Given $a_2 a_3 a_4 a_5$ and T, a_1 is determined.

Let n_5 be the number of *left* $n5$ template positions. Then the probability that $P_1 P_2 P_3 P_4 P_5$ belongs to a neighbourhood on an $n5$ template equals $n_5/16$. If so, $P_1 P_2 P_3 P_4$ and I_3 will determine the *next cell* P_5, as illustrated in the diagram below (conversely, *right*).

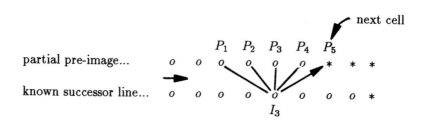

3.6 The Z Parameter

n=4. The *n4 template* (left or right), size 4, has eight possible positions and occupies 1/8 the rule table.

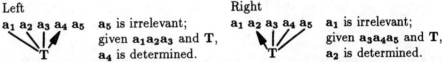

The underline (or linked underlines) indicates entries are equal; the curved link indicates the two sets of equal entries are different.

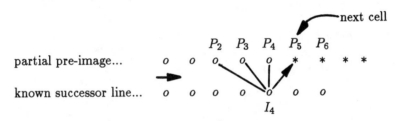

Let n_4 be the number of *left* n4 template positions. Then the probability that $P_2P_3P_4P_5P_6$ belongs to a neighbourhood on an n4 template equals $n_4/8$. If so, $P_2P_3P_4$ and I_4 will determine the *next cell* P_5, as illustrated in the diagram below (conversely, *right*).

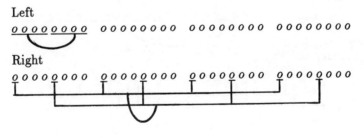

n=3. The *n3 template* (left or right), size 8, has four possible positions and occupies 1/4 the rule table.

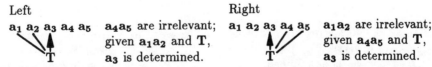

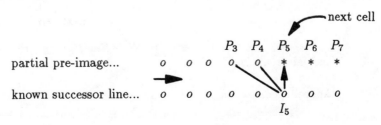

Let n_3 be the number of *left* n3 template positions. Then the probability that $P_3P_4P_5P_6P_7$ belongs to a neighbourhood on an n2 template equals $n_3/4$. If so, P_3P_4 and I_5 will determine the *next cell* P_5, as illustrated in the diagram below (conversely, *right*).

n=2. The *n2 template* (left or right), size 16, has two possible positions and occupies 1/2 the rule table.

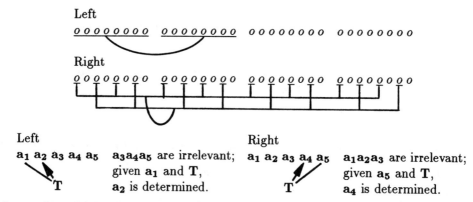

Let n_2 be the number of *left* n2 template positions. Then the probability that $P_4P_5P_6P_7P_8$ belongs to a neighbourhood on an n2 template equals $n_2/2$. If so, P_4 and I_6 will determine the *next cell* P_5, as illustrated in the diagram below (conversely, *right*).

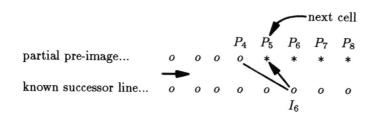

n=1. The trivial n1 *template* (left or right), size 32, has one possible position and occupies the entire the rule table.

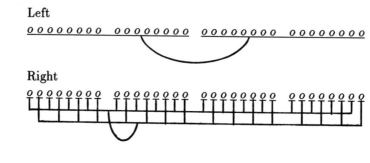

If a *left* n1 template fits, then $n_1 = 1$. The probability that $P_5P_6P_7P_8P_9$ belongs to a neighbourhood on an n1 template equals n_1. If so, I_7 will determine the *next cell* P_5, as illustrated in the diagram below (conversely, *right*).

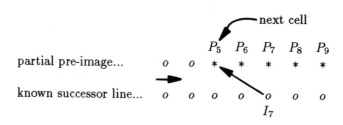

3.6 The Z Parameter

To summarise, given an $n = 5$ rule, the number of positions where the five types of template fit the rule table are designated as follows:

$$n_5 \text{ (max 16)} \ldots n5 \text{ template (size 2)}$$
$$n_4 \text{ (max 8)} \ldots n4 \text{ template (size 4)}$$
$$n_3 \text{ (max 4)} \ldots n3 \text{ template (size 8)}$$
$$n_2 \text{ (max 2)} \ldots n2 \text{ template (size 16)}$$
$$n_1 \text{ (max 1)} \ldots n1 \text{ template (size 32)}$$

Templates relating to a given direction are mutually exclusive; a given rule table entry may not connect to more than one left template and one right template. The following method will give the corrected probability that the *next cell* is determined, thus the corrected value of Z. The method described below computes Z_L, from left to right. A converse but otherwise identical method computes Z_R, from right to left.

- $n5$: let p_5 be the probability that the *next cell* is determined by $n5$, $p_5 = n_5/16$; the probability that it is *not* determined, $\overline{p}_5 = 1 - p_5$.
- $n4$: if *not* determined by $n5$, let p_4 be the probability that the *next cell* is determined by $n4$, $p_4 = n_4/8 \times \overline{p}_5$; the probability that it is *not* determined, $\overline{p}_4 = 1 - p_5 - p_4$.
- $n3$: if *not* determined by $n4$, let p_3 be the probability that the *next cell* is determined by $n3$, $p_3 = n_3/4 \times \overline{p}_4$; the probability that it is *not* determined, $\overline{p}_3 = 1 - p_5 - p_4 - p_3$.
- $n2$: if *not* determined by $n3$, let p_2 be the probability that the *next cell* is determined by $n2$, $p_2 = n_2/2 \times \overline{p}_3$.
- $n1$: this is a special case, as the $n1$ template takes up the whole rule table. Let p_1 be the probability that the *next cell* is determined by $n1$, $p_1 = n_1$. (Note: if $n_1 = 1$, Z_L and $Z_R = 1$)

$$Z_L = p_5 + p_4 + p_3 + p_2 + p_1.$$

The corrected parameter $Z = Z_L$ or Z_R, whichever is greater.

As an example, consider the $n = 5$ rule 2334561936, which has a space-time pattern illustrated in Fig. 2.2. The rule table is set out below:

Left

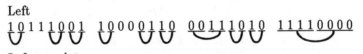

Left templates:
$n_5 = 8, \quad n_4 = 1, \quad n_3 = 1$

Right

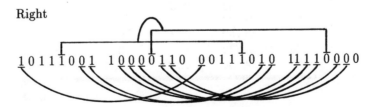

Right templates:
$n_5 = 8, \quad n_4 = 1$

Left procedure

$p_5 = n_5/16 = 8/16 = .5$ $\qquad$ $\overline{p}_5 = 1 - .5 = .5$
$p_4 = n_4/8 \times \overline{p}_5 = 1/8 \times .5 = .0625$ $\qquad$ $\overline{p}_4 = 1 - .5 - .0625 = .4375$
$p_3 = n_3/4 \times \overline{p}_4 = 1/4 \times .4375 = .109375$

$$Z_L = p_5 + p_4 + p_3 = .5 + .0625 + .109375 = .671875$$

Right procedure

$$p_5 = n_5/16 = 8/16 = .5 \qquad \overline{p}_5 = 1 - .5 = .5$$
$$p_4 = n_4/8 \times \overline{p}_5 = 1/8 \times .5 = .0625$$

$$Z_R = p_5 + p_4 = .5 + .0625+ = .5625$$
$$Z_L > Z_R, \text{ so } Z = .671875$$

The method of computing Z gives the same result for an $n = z$ rule expressed as an $n > z$ rule. For instance, the $n = 3$ rule 54, 0011-0110, may be expressed as the $n = 5$ rule 255594300, 0000111100111100-0000111100111100. Z is the same for both rule tables.

Z is equal for rules belonging to the same rule cluster, and $Z_L = Z_R$ for *symmetric* and *fully asymmetric* rules. $Z_L \neq Z_R$ for *semi-asymmetric* rules.

The Z parameter could be extended for rules with greater neighbourhood size n, and generalised for greater value range k. The relationship between Z and Langton's λ parameter[16,17] is discussed in chapter 4.

A set of tables, 3.2 and 3.3, showing the λ parameter, λ ratio (see chapter 4), and Z parameter for $n = 3$ rules and $n = 5$ totalistic codes appears below. Only the lowest rule number or code in each cluster is listed, and represents the whole cluster.

TABLE 3.2 $n = 5$ totalistic codes

rule code	λ parameter	λ ratio	Z parameter
0	0/32	0	0
1	1/32	0.0625	0.0625
2	5/32	0.3135	0.3125
3	6/32	0.375	0.25
4	10/32	0.625	0.625
5	11/32	0.6875	0.6875
6	15/32	0.9375	0.4375
7	16/32	1	0.375
9	11/32	0.6875	0.6875
10	15/32	0.9375	0.9375
11	16/32	1	0.875
12	20/32	0.75	0.5
13	21/32	0.6875	0.5625
14	25/32	0.4375	0.3125
17	6/32	0.375	0.375
18	10/32	0.625	0.625
21	16/32	1	1
22	20/32	0.75	0.75
25	16/32	1	0.5
30	30/32	0.125	0.125

3.6 The Z Parameter

TABLE 3.3 $n = 3$ rules

	rule number	λ parameter	λ ratio	Z parameter
symmetric rules	0	0/8	0	0
	1	1/8	0.25	0.25
	4	1/8	0.25	0.25
	5	2/8	0.5	0.5
	18	2/8	0.5	0.5
	19	3/8	0.75	0.625
	22	3/8	0.75	0.75
	23	4/8	1	0.5
	33	2/8	0.5	0.5
	36	2/8	0.5	0.5
	37	3/8	0.75	0.75
	50	3/8	0.75	0.625
	51	4/8	1	1
	54	4/8	1	0.75
	73	3/8	0.75	0.75
	77	4/8	1	0.5
	90	4/8	1	1
	94	5/8	0.75	0.75
	105	4/8	1	1
	126	6/8	0.5	0.5
semi-asymmetric rules	2	1/8	0.25	0.25
	3	2/8	0.5	0.5
	6	2/8	0.5	0.5
	7	3/8	0.75	0.75
	9	2/8	0.5	0.5
	12	2/8	0.5	0.5
	13	3/8	0.75	0.75
	26	3/8	0.75	0.75
	27	4/8	1	0.75
	30	4/8	1	1
	35	3/8	0.75	0.625
	38	3/8	0.75	0.75
	41	3/8	0.75	0.75
	45	4/8	1	1
	58	4/8	1	0.75
	62	5/8	0.75	0.75
fully asymmetric rules	10	2/8	0.5	0.5
	11	3/8	0.75	0.75
	14	3/8	0.75	0.75
	15	4/8	1	1
	24	2/8	0.5	0.5
	25	3/8	0.75	0.75
	28	3/8	0.75	0.75
	29	4/8	1	0.5
	43	4/8	1	0.5
	46	4/8	1	0.5
	57	4/8	1	0.75
	60	4/8	1	1

FOUR

Implications of Basin of Attraction Fields

4.1 Basin Field Topology and Rule Space

In this section, some implications of the emerging basin field landscape on the current perception of the structure of rule space are examined. Wolfram has proposed that all CA rules belong to one of four universality classes.[34,36,39] These classes are essentially phenomenological,[8] based on the characteristic appearance of typical space-time patterns. Wolfram's classes, and their analogues in continuous dynamical systems, are said to exhibit the behaviour shown below.

CA behaviour	dynamical systems analogue
class 1 evolves to a fixed, homogeneous state	limit points
class 2 evolves to separated periodic regions	limit cycles
class 3 evolves to *chaotic*, aperiodic patterns	strange attractors
class 4 evolves to *complex*, localised structures	long transients, no analogue

It is accepted that many rules show "intermediate" behaviour,[36] and there is also scope for defining sub-classes,[19,20] but the discussion below uses a loose definition of the four classes listed. Of these classes, it is conjectured that the supposedly rare[16] *complex rules* (class 4) may in some cases be capable of supporting *universal computation*.[17,34] The space-time patterns of the complex rules contain interacting, static, and propagating structures, sometimes called *information structures*, similar to those illustrated in Figs. 2.2 and 4.1.

Langton proposed that class 4 rules are located at a *phase transition* in rule space. He suggested that the long transients typical of these rules have potential for information processing, and implications for understanding the origin and evolution of life.[17]

4.1.1 The λ Parameter

Rule space has been characterised by the λ parameter,[16,17] the proportion of *non-zero* entries in the rule table. If the value range is k, with possible values $0, 1, 2, \ldots, k-1$, then a particular value, say 0, is selected as the *quiescent* value. The λ parameter is the proportion of rule table entries other than 0.

Langton and others[17,20,41] have shown that the rule classes can be roughly selected by adjusting λ between 0 and $1 - (1/k)$. A rule table may be constructed by assigning one of the k values to each entry with equal probability, so that the density of all values, including the quiescent value, will be roughly equal, and $\lambda \simeq 1 - (1/k)$. At this value of λ, space-time patterns are most likely to appear chaotic.

Rules may be selected according to other values of λ. This is typically done by assigning the quiescent value with a selected probability x, so that $\lambda = 1 - x$. The remaining values are then assigned with equal probability $\lambda/(k-1)$.

For binary rules where $k = 2$, λ is simply the density of 1s in the rule table. Maximum chaos in the appearance of space-time patterns is likely to occur at $\lambda = 1/2$, a rule table with an equal number of 0s and 1s.

As the λ parameter is varied from 0 to 1/2, the various classes of behaviour are traversed. Complex, class 4 behaviour occurs at a *phase transition* between periodic, class 2 and chaotic, class 3 behaviour, reordering Wolfram's classes as follows[16,17]:

$$
\begin{array}{rcccccccc}
 & & & & & \text{complex} & & & \\
 & & & & & \text{rules} & & & \\
 & & & & & | & & & \\
\text{class}: & 1 & \longrightarrow & 2 & \longrightarrow & 4 & \longrightarrow & & 3 \\
\text{in general}, \lambda: & 0 & \ldots & \ldots & \ldots & \ldots & \ldots & & 1 - (1/k) \\
\text{binary}, k = 2, \lambda: & 0 & \ldots & \ldots & \ldots & \ldots & \ldots & & 1/2
\end{array}
$$

As λ increases from 1/2 to 1 the sequence is reversed, so that the complex rules occur in two limited regions on either side of $\lambda = 1/2$.

4.1.2 Convergence of State Space

It is suggested that the λ parameter is modulating the same aspect of CA global behaviour as the Z parameter (introduced in chapter 3, section 3.6)—the *maximum pre-imaging* (mp) as a function of array length. In general this will reflect the *degree of pre-imaging* typical of a rule. The degree of pre-imaging is apparent in the topology of a rule's basin of attraction field, and reflects the *convergence of state space*,[14] which may be equivalently measured as the density of garden-of-Eden nodes in state space (or in one basin).

Low pre-imaging implies low convergence and a low density of garden-of-Eden nodes; high pre-imaging implies high convergence and high density of garden-of-Eden nodes. The degree of pre-imaging and the density of garden-of-Eden nodes are available in the Atlas (or program).

4.1.3 The Z Parameter and the λ Parameter

The Z parameter is the probability that the *next cell* of a partial pre-image has a unique value (see chapter 3, section 3.6). For binary rules, when $Z = 1$, the quantity of 0s and 1s in the rule table is equal, $\lambda = 1/2$; when $Z = 0$, $\lambda = 0$. As Z varies between 0 and 1, λ varies between 0 and 1/2 or conversely between 1 and 1/2.

The Z parameter may possibly be the mechanism underlying the operation of the λ parameter, because the probability of a rule table having a given value of Z depends on the proportions of 0s and 1s assigned at random to the rule table according to the setting of λ. Thus λ seems to be a measure of the *probability* of a particular value of Z. The Z parameter is concerned not only with the numbers of 0s and 1s in the rule table, but also their position and thus might be expected to modulate behaviour more closely.

For a more direct comparison between the values of λ and Z for a given binary rule table, the λ parameter may be modified as follows. The quiescent state is taken as the majority value in the rule table; the minority value is taken as the active state. The modified λ parameter, referred to as the λ *ratio*, is the ratio of the active values to 1/2 of the rule table (the potential maximum of active values). For example, in an $n = 5$ rule table with 32 entries, say 10 entries in the rule table are 0, then the λ ratio equals 10/16.

$$\text{If } \lambda \leq 1/2, \text{ then } \lambda \text{ ratio } = 2\lambda.$$
$$\text{If } \lambda > 1/2, \text{ then } \lambda \text{ ratio } = 2(1 - \lambda).$$

The value of the λ ratio varies between 0 and 1 roughly in line with, but never smaller than, the Z parameter. $Z \leq \lambda$ ratio, and $Z = \lambda$ ratio $= 1$ only for limited pre-image rules.

As an example of the λ ratio and the Z parameter, consider the space-time patterns with complex structures that were illustrated in Figs. 2.2(a–c). The relevant rules have the following values for the λ ratio and Z:

a. rule 3112581872 λ ratio $= 1$, $Z = .671875$
b. rule 2334561936 λ ratio $= .8125$, $Z = .6875$
c. rule 3583552890 λ ratio $= .875$, $Z = .75$

4.1 Basin Field Topology and Rule Space

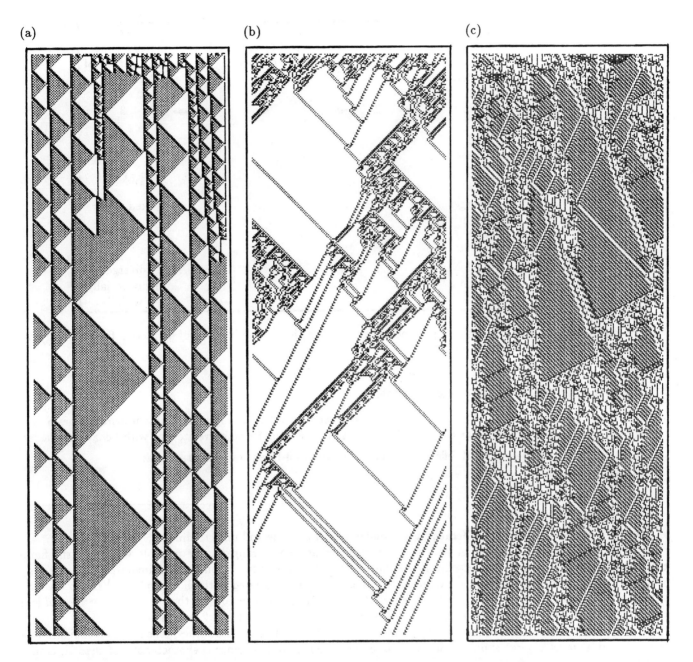

FIGURE 4.1 Space-time patterns for $n = 5$ rules. Array size is 150, 440 time steps from a random initial state. Rule numbers, λ ratio, and Z parameter are: (a) rule 1598319856, λ ratio = 1, $Z = .58984375$. (b) rule 1550470552, λ ratio = .9375, $Z = .7265625$. (c) rule 2912711218, λ ratio = 1, $Z = .75$.

There are many examples of rules where the λ ratio equals 1 or close to 1, suggesting a chaotic space-time pattern, when in fact the pattern appears complex. In such cases the value of Z will typically be between .6 and .8. In Figure 4.1 are some examples of space-time patterns for various $n = 5$ rules that appear to belong to class 4, with complex interacting large-scale structures. The values of λ ratio and Z are shown.

	HIGH convergence and garden of Eden density		phase transition mp and (mc, ml, nb) in balance		LOW convergence and garden of Eden density
class	1 →	2 →	4	→	3
Z:	0	... ...	...	...	1
λ ratio:	0	... ...	...	...	1
λ:	0	... ...	...	...	$1 - (1/k)$
mp:	diverges exponentially with L	... ...	...	...	fixed upper limit
(mc, ml, nb):	fixed lower limit	... ...	...	...	diverges exponentially with L

FIGURE 4.2 Relationship between the Z parameter (and λ) and the change in basin of attraction field topology as L increases, reflecting convergence of state space.

4.1.4 Basin Field Topology and the Z Parameter

As Z is varied between 0 and 1, it may be possible to predict how the following quantifiable features of basin field topology will vary with increasing array length L (see data in the Atlas), and correlate this with behaviour classes.

1. mp, maximum pre-imaging (will reflect the density of garden-of-Eden states in state space).
2. mc, maximum period of attractor cycles.
3. ml, maximum length of transient trees.[1]
4. nb, the number of separate basins in the field.

mc, ml, and nb will vary together, because a preponderance of any one will tend to diminish the other two.

Taking the general case of a CA with neighbourhood n and value range k, the possible ways that the finite number of states, k^n, can be connected together into basins depends critically on the maximum pre-imaging, mp. Kauffman has observed that "high convergence in state space...is associated with short cycles."[14] If mp diverges exponentially with L, ($Z < 1/2$), most states will be locked up in pre-image fans, so that the scope for mc, ml and nb to diverge with L will be severely limited. This is characteristic of class 1 and 2 behaviour.

Conversely, if mp is held constant irrespective of L ($Z = 1$, limited pre-image rules), then some combination of mc, ml, and nb must diverge exponentially with L. Such exponential divergence is characteristic of chaotic, class 3 behaviour.

At an intermediate value of Z, mp and some combination of mc, ml, and nb will be finely balanced and will diverge by some intermediate function with L. An example of an intermediate function for a binary rule is $\sqrt{2^L}$, and in general $(k^L)^{1/k}$.

---

[1]In this paper, the length of transient trees is interpreted literally as the number of time steps from the furthest garden-of-Eden node to the attractor. However, beyond the "phase transition," the space-time patterns may quickly settle to what appears to be a "chaotic steady state." The number of time steps to reach this chaotic state (typically becoming *shorter* with increasing λ) is an alternative, descriptive interpretation of transient length, for instance.[17]

Following our literal interpretation, transient length typically continues to *increases exponentially* to an ever greater degree after the phase transition. Although the characteristic appearance of the space-time patterns, whether on a long transient or in the attractor, may be indistinguishable, the difference is clear in a state transition diagram.

4.1 Basin Field Topology and Rule Space

It has been suggested that as the λ parameter is increased, this balancing point will emerge abruptly, and locates the class 4 complex rules at a *phase transition* in rule space.[17]

It is evident from the atlas that there is a certain trade-off between mc, ml, and nb, as Z approaches 1, and that often only one or two of these variables will diverge with L. For example, consider the complementary $n = 3$ limited pre-image rules, 45 and 210, where $Z = 1$, and $mp = 3$ irrespective of L. In rule 45, mc and ml diverge exponentially with L; nb diverges to a limited extent. By contrast, in rule 210, mc and ml diverge to a limited extent with L, whereas nb diverges exponentially. The mechanism of the trade-off between mc, ml and nb is unclear.

Figure 4.2 summarises the expected variation of the characteristics of the basin field with array size L, as Z is varied between 0 and 1, and the corresponding rule classes.

4.1.5 Examples of Typical Basin Topology in Relation to Rule Class

To illustrate the relationships in Fig. 4.2, an example of a typical basin of attraction for $n = 3$ rules characteristic of each rule class is given in Figs. 4.3–4.6, including the following data: mp, maximum pre-imaging; *g-density*, density of garden-of-Eden nodes; *period*, of attractor cycle; and ml, maximum transient length.

In these examples, rotation equivalent transients are suppressed for the sake of clarity. General examples of the complete basin field over a range of array length L are presented in the Atlas.

CLASS 1. Rules where $Z = .25$ will have a large proportion of ambiguous permutations in the rule table, so mp will diverge exponentially with L; mc, ml, and nb will either remain fixed or relate arithmetically to L. Most states are locked into pre-image fans flowing into point attractors, or two-state attractors, especially the *uniform* states (all 0s or 1s). The few states outside the influence of these attractors will only have enough scope to form basins with short cycles and transients. (See Fig. 4.3.)

CLASS 2. Rules where $Z = .5$ may have many basins separated from the uniform attractor. mp will diverge exponentially with L, but to a lesser extent than for class 1. mc, ml, and nb will relate arithmetically to L. Again many states are locked up in pre-image fans leaving enough scope for only short cycles or transients. (See Fig. 4.4.)

CLASS 4. The complex, class 4 rules occur typically at $Z = .75$. mp and some combination of mc, ml, and nb will be finely balanced and will diverge by some intermediate function with L. Controlled pre-imaging allows enough scope for moderately long cycles and transients. (See Figure 4.5.)

CLASS 3. The chaotic, class 3 rules are limited pre-image rules where $Z = 1$. mp is fixed irrespective of L. Some combination of mc, ml, and nb diverge exponentially with increasing L. For instance, clusters 30 and 45, where $mp = 3$, have the greatest values of mc and ml among the $n = 3$ rules. (See Fig. 4.6.)

In general, the space-time patterns of the $n = 3$ and $n = 5$ rules have been found to correlate with the Z parameter; however, as with the λ parameter, there are exceptions. It has been noted that the transition to chaotic behaviour may occur at different λ values, although there is a well-defined distribution around a mean value.[17,19] Chaotic space-time patterns may also occur at low values of Z. For example, the $n = 3$ rules 18 and 126 have a value of $Z = .5$. Although the typical basin topology of these rules correlates with Z, their space-time patterns appear chaotic, but are confined to only a subset of possible configurations of neighbourhoods. This is clearly seen when colors are assigned to cells in space-time patterns according to the neighbourhood which determined the cell's value, implemented in the program Space1 following a suggestion by Warrell.[43] The process whereby active neighbourhoods are eliminated requires further investigation.

The space-time patterns of the class 2, 4, and 3 rules used in the examples are shown in Figure 4.7.

FIGURE 4.3 Class 1, rule 251. $L = 12$, seed 111111111111. $mp = 853$, $g\text{-}density = .92$, period $= 1$, $ml = 6$. mp diverges exponentially with L; $Z = .25$; (λ ratio $= .25$). Equivalent transients above level 1 are suppressed.

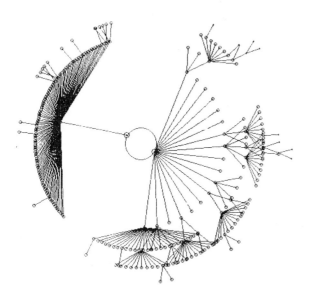

FIGURE 4.4 Class 2, rule 33. $L = 16$, seed 0111111111100000. $mp = 97$, $g\text{-}density = .84$, period $= 2$, $ml = 5$. mp diverges exponentially with L; $Z = .5$; (λ ratio $= .5$).

4.2 Mutation

Kauffman and others have made the analogy between CA behaviour and biological systems.[14,15,20] The rule table is regarded as the *genotype* and the dynamics shown by the rule, the *phenotype*. The rule, as expressed by its rule table, is compared to a DNA sequence. Experiments may be done to see how changing, or mutating, the elements in a rule table may change behaviour in terms of typical space-time patterns.

The same investigation may now be applied to basin of attraction fields, which in principle represent all possible space-time patterns or global behaviour. The effects of a mutation of the rule table on the topology, or form, of the basin field has added significance for biology because it is analogous to mutation of the genetic code altering the morphological form of an organism.

The *Hamming distance* between two rules specifies the extent of a mutation of a binary rule table, and is a measure of the number elements, or bits, in the rule table that have been changed. A Hamming distance of one bit is the smallest possible mutation of a given rule table.

4.2.1 Mutant Basins of Attraction

A binary rule table has 2^n entries. A given rule will have a set of 2^n mutants separated from it by one bit. In the

4.2 Mutation

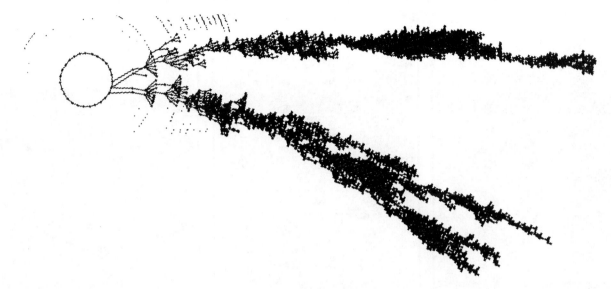

FIGURE 4.5 Class 4, rule 193. $L = 18$, seed 011010001110000010. $mp = 70$, $g\text{-}density = .61$, period $= 27$, $ml = 120$. mp, mc, and ml diverge with L; $Z = .75$; (λ ratio $= .75$). Equivalent transients are suppressed.

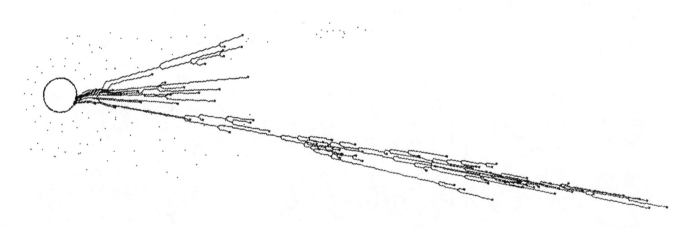

FIGURE 4.6 Class 3, rule 30. $L = 15$, seed 110110111000000. $mp = 2$, $g\text{-}density = .04$, period $= 1455$, $ml = 321$. mp and ml diverge exponentially with L; $Z = 1$; (λ ratio $= 1$). Equivalent transients and pre-image nodes are suppressed; angle between pre-image arcs is increased.

case of the $n = 3$ rule table, each rule will have 8 one-bit mutants. An $n = 5$ rule table will have 32 one-bit mutants.

In the $n = 3$ rule table, with eight entries, the smallest mutation of the rule table, one bit, is likely to result in a very significant change to the basin field topology, because a relatively large change, 12.5%, has been made to the rule table.

As the size of the rule table, 2^n, is increased, the rule space becomes "smoother." The significance of the smallest mutation decreases. Mutating the $n = 5$ rule table, with 32 entries, by one bit, results in a relatively smaller change to the rule table, 3.125%, and a correspondingly less significant change in basin field topology. The set of 32 one-bit mutant basin topologies are recognisably related. If further mutations are made to the rule table, the divergence increases.

It is possible to mutate an $n = 3$ rule by a change of less than 12.5%, if it is expressed as an $n = 5$ rule (see chapter 3, section 3.3.9). Mutating the $n = 5$ rule table by one bit means moving in rule space away from the $n = 3$ rule into the space of $n = 5$ rules that are as close as possible to the $n = 3$ source rule. Rules in the $n = 3$ rule

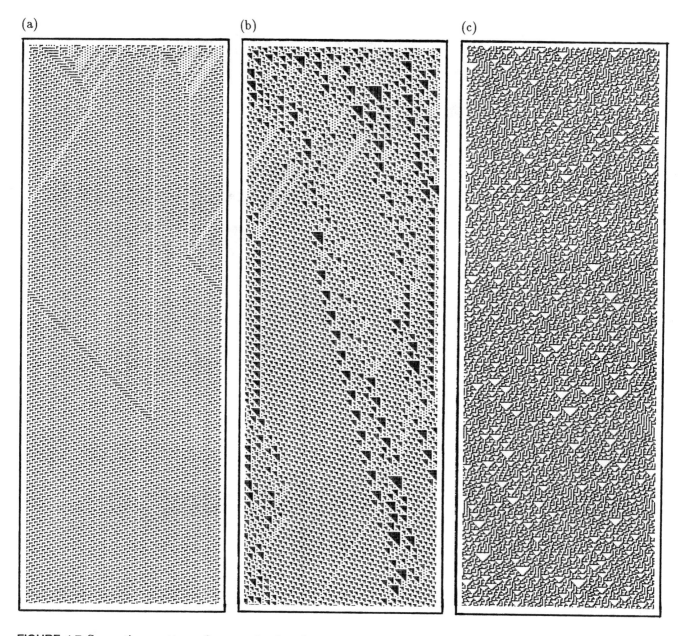

FIGURE 4.7 Space-time patterns for $n = 3$ rules. Array size is 150, 420 time steps from a random initial state. Rule numbers and Z parameter are: (a) Class 2: rule 33, $Z = .5$; (b) Class 4: rule 193, $Z = .75$; (c) Class 3: rule 30, $Z = 1$.

subset, far from forming a cohesive group, are scattered within $n = 5$ rule-space. In the same way, $n = 5$ rules are scattered within $n = 7$ rule space, and so on.

Fig. 4.8 shows the basin field for the $n = 3$ rule 195, $L = 8$ (the source rule), located in the lower left corner of the diagram. In fact, the basin field consists of a single basin, with a point attractor whose global state is all 1s. The other 32 panels in the diagram show the single basin of attraction containing the seed all 1s, according to a rule mutated by one bit from the $n = 5$ expression of the source rule. The mutations are made to successive bits in the rule table from left to right, numbered 1 to 32. The corresponding one-bit mutant basins are located in ascending rows from left to right.

4.2 Mutation

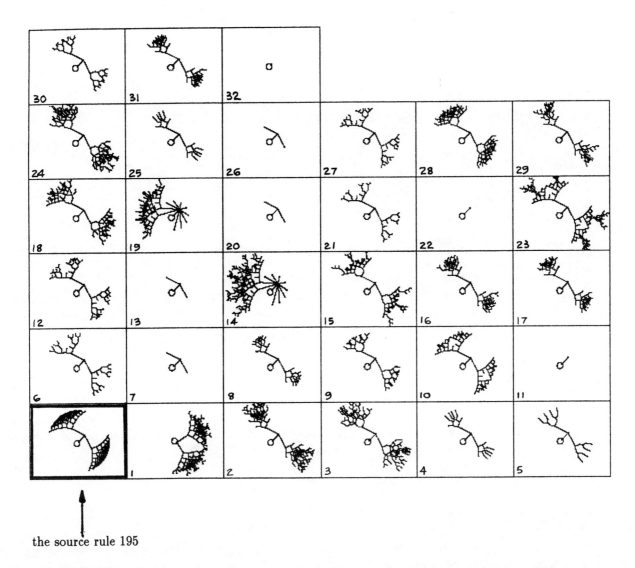

the source rule 195

FIGURE 4.8 Rule 195 and 32 one-bit mutants. Seed state—all 1s; $L = 8$. The source rule is at the lower left corner. Successive mutants (1–32) are in ascending rows from left to right (angle between pre-image arcs increased).

In general, $n = 5$ rule space has been found to have a measure of continuity from a subjective appraisal of basin field topology, but sharp changes are also present. Rule space for $n > 5$, with a larger rule table, would be expected to be proportionately smoother.

Rules may be mutated cumulatively by 32 one-bit steps, until the rule table is turned into its complement. Appendix 4 shows further examples of sets of one-bit mutants and cumulative mutants.

The program allows automatic mutation of rules (see Appendix 1) to produce mutated basin of attraction fields (or single basins), or mutated space-time patterns. Sets of 32 mutants as illustrated may be assembled. Space-time patterns may be mutated by one bit (and mutated back) in mid run to allow the evolution and selection of a preferred pattern. Complex space-time patterns may be easily found in this way.

4.2.2 Mated Rules

Rules may be mated by combining the left half of one rule table with the right half of another. If the "parent" rule tables are very different, having different basin structure, then the *offspring* will have an intermediate basin

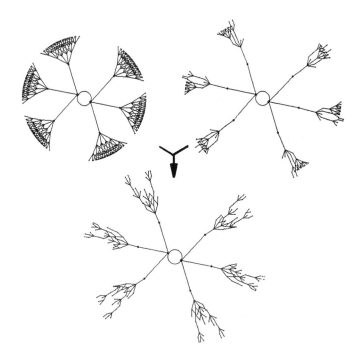

FIGURE 4.9 The $n = 3$ parent rules 105, 01101001, and 73, 01001001, and their offspring, the $n = 5$ rule 818101443, 0011110011000011-0011000011000011, L = 12, seed all 0s.

structure. If the parents are separated by only one bit, then mating rules will result in an offspring that is identical to either one parent rule or the other. Slightly greater distance between parent rules results in greater variation, but with the offspring basin structure bearing a closer family resemblance to each parent than the parents to each other. The parallels with sexual reproduction are clear. An example is given in Fig. 4.9.

4.3 Conclusion

This book demonstrates the complex unfolding of basin of attraction fields, for CA constructed with simple parameters. This complements, in terms of global behaviour, the well-documented space-time patterns of individual trajectories. Access to basin field structure may allow insights into the global behaviour of CA with more complex parameters, and indeed to any dynamical model or real system that evolves within a basin of attraction field. The basin field is self-organised according to a linear code, and responds to mutations of the code, a poignant analogue to genetic systems.

A synoptic view of basin structure has suggested a number of principles underlying CA behaviour and the structure of rule space. It has been shown that rotational symmetry, and bilateral symmetry for symmetrical rules, are conserved in CA evolution. The organisation of rule equivalence classes into rule clusters and symmetry categories has been proposed. A general reverse algorithm for generating all pre-images to a given CA state has been presented.

The subset of limited pre-image rules has been identified, and the mechanism within a rule table that modulates the degree of pre-imaging has been quantified by a proposed Z parameter. It has been shown that the degree of pre-imaging is likely to be a key variable controlling the topology of basin of attraction fields, and may explain the operation of the λ parameter used to classify rule space.[17]

The raw data in the atlas, and the program for generating basin fields, will, it is hoped, provide further avenues of research.

APPENDIX 1

The Atlas Program

The programs included with this volume are a research tool under development, and not presented as a consumer product. The programs run only on PC-compatible computers in DOS; Apple Macintosh computers can be made to emulate DOS, but the authors have not experimented with what kind of hardware would be required. The software is purchased "as is." Those interested in receiving future developments of the software (including versions for the Apple Macintosh and Sun Microsystems computers) should write to: Mr. A. Wuensche, 48 Esmond Road, London W4 1JQ, U.K. For the latest features update, run the "update" file included with the diskette.

The authors retain copyright on the software and the images implicit in the software, which may not be published without the authors' permission. Copyright © 1990 A.Wuensche & M.Lesser. All rights reserved.

A1.1 Operating Instructions

A1.1.1 Hardware/Software Recommendations/Requirements

- *IBM PC or compatible, with DOS 2.0 or above
- 80286 chip or above, math co-processor recommended
- 640k memory available
- *VGA (640 × 480) 16-colour graphics card
- Epson MX-80 dot matrix printer, or compatible

  Variation from the above may cause problems. Items marked "*" are essential.

A1.1.2 The Programs

There are three programs on a 3.5-inch diskette enclosed inside the back cover:

- ATLAS1.EXE: for drawing basin of attraction fields.
- SPACE1.EXE: for drawing space-time patterns.
- UPDATE.EXE: for the latest features not described in the operating instructions.

A1.1.3 The Setup and Graphics Screens and the Interrupt Key (F12)

The Atlas1 and Space 1 programs divide into two parts, the setup screen where parameters are specified and the graphics screen where the graphics are drawn. In both screens, except when printing, loading, or saving, **key F12 + RETURN** will interrupt operation and allow options as follows:

The interrupt key (F12)
- `abort`   To the beginning of the setup screen
- `dump`   The screen image to the printer
- `continue`   From point of interrupt

for ATLAS1 only
- `save`   The screen image to disk
- `load`   A screen image file, or series of files, from disk

for SPACE1 only
- `revise`   The revision menu

key (F11) will interrupt while printing

At the foot the screen, the following prompt and title appears:

F12-interrupt   Networks of Attraction/copyright © A.Wuensche & M.Lesser/May 1990.

or

F12-interrupt   Space-Time Patterns/copyright © A.Wuensche & M.Lesser/May 1990.

A1.2 ATLAS1: Basin of Attraction Fields

A1.2.1 Running the Program; the Set-up Screen

To run: insert the diskette into a drive, or save onto hard disk; if in drive A, at the DOS prompt, enter `A:ATLAS1`. The setup screen (grey background) will appear, with this message:

```
free date bytes 460048 (or similar), at least 180000 required.
to quit Atlas1 enter 99 + RETURN, else RETURN:
```

The following is a brief commentary on each setup choice. Press RETURN for the next choice.

In general, 99 selects an option, and RETURN the default. A series of prompts are presented to specify the rule parameters, graphics screen layout, and the seed line. (Note that NAT, short for Network of ATtraction, is used to denote a basin of attraction/state transition diagram.)

A1.2.2 Select Either Single NAT (or transient branch only) or NAT Field

A single NAT is one basin in the basin field, whereas the NAT field is the entire set of nonequivalent basins making up the basin field.

```
for single NAT enter 99, else for NAT field RETURN:
transient branch only enter 99, else for complete NAT RETURN:
```

A transient branch is a fragment of the NAT consisting of all upstream states from a given seed state. Note that if the seed state is on the attractor, the sequence of upstream states will be endless.

A1.2.3 NAT Information to the Printer

Selects automatic printing of NAT data (see A1.2.14) as NATs are being drawn. Selecting this option when the printer is not ready will cause "error - 57 - printer not ready."

```
for NAT info to printer enter 99, else RETURN:
```

A1.2.4 Select the Line Length or Range of Line Lengths

A single NAT or the NAT field may be drawn for either one particular line length, or over a range of line lengths. For a range of line lengths, specify the first and last line lengths; for one line length only, specify the same line length for first and last. There are different defaults according to *single NAT* or *NAT field* selected in A1.2.2. For instance, for a NAT field,

```
select first line length, (default 1), 1-18:
select last line length, (default 10), 1-18:
```

The maximum line length for a *single NAT* is 31 and for a *NAT field*, 18. The minimum line length is 1. "error - 14 - out of memory" may occur, if the configuration of rule and seed, combined with an excessive line length, results in:

a. too great a *disclosure length*,

b. too many *partial pre-images*, or

c. too many *pre-images* at the same level on one transient tree.

Note that the time taken to draw a NAT field tends to increase as 2^{L-1}, where L is the line length, rather than 2^L. This is because equivalent basins, equivalent transient trees, and equivalent transient branches from uniform states do not need to be regenerated.

A1.2.5 Select the Type of Rule and Notation

The rule type may be selected as an $n = 5$ rule (in decimal or binary), an $n = 3$ rule (the default), or an $n = 5$ totalistic code (the specific rule will be chosen in A1.2.7). There is also the choice of a random $n = 5$ rule.

```
Select notation to specify 5-rule or subset
1)  5-rule Decimal
2)  5-rule Binary
3)  3-rule (default)
4)  5-rule Totalistic code
5)  5-rule Random
enter 1, 2, 3, 4, 5 or RETURN:
```

A1.2.6 Specify Automatic Mutation, Screen Save, or Printer Dump

When the graphic image is complete, RETURN will result in a new graphic image according to a mutated rule, retaining the current setup. The default is a one-bit mutation of the source rule, starting from the leftmost position in the rule table. After 32 mutations you are returned to the setup screen.

Mutations are made to the $n = 5$ expression of totalistic codes and $n = 3$ rules. A random $n = 5$ next rule may also be selected. Mutations may be automatically saved, or printed (see A1.2.17). Selecting "next rule without pause" and option 1 or 4 results in an endless succession of graphic output.

```
             to specify mutated or automatic next rule
             or automatic screen save or printer dump,
             enter 99, else RETURN:
```

If 99 is selected, the various options for mutation and automatic saving or printing are presented:

```
             1)  increment rule number by +1
             2)  32 single mutations from left (default)
             3)  32 cumulative mutations from left
             4)  a new rule selected at random
                 enter 1, 2, 3, 4 or RETURN:
```

If option 2 or 3 is selected, the position of the first bit to be changed, from the left, may be specified. The default is 1, i.e., the leftmost bit.

```
             enter start mutation 1-32 (default = 1):
```

NATs for successive rules may be drawn without a pause:

```
             next rule without pause enter 1, else RETURN
```

Each complete screen for successive rules may be automatically saved. This permits a set of 32 mutant NATs to be assembled (see A1.2.17).

```
             for automatic screensave enter 1, else RETURN:

             enter mutant filename max 6 characters
             default is d:myfile (? to list):
```

Each complete screen for successive rules may be automatically dumped to the printer, with an optional pause before each dump:

```
             for automatic printer dump enter 1, else RETURN:
             to pause before dump enter 1, else RETURN:
```

A1.2.7 Select the Rule

According to the choice of rule type and notation in A1.2.5, the rule is selected in decimal or binary (the default is a random rule). For example,

```
             select 3-rule notation 1) decimal (default), 2) binary
             enter 1, 2 or RETURN:

             enter 3-rule number 0 to 255 (default random):
             or
             enter 0s and 1s to make up an 8-bit binary number
```

A1.2.8 Reposition or Rotate

The centre of the first NAT may be located anywhere on (or off) the screen. The screen is centrally divided by notional x, y coordinates. The centre of the screen is at $x = 0, y = 0$, the left and right edges at $x = -100\%$ and $+100\%$, and the top and bottom edges at $y = +100\%$ and -100%.

Defaults for the initial position vary according to parameters previously selected. The position may be specified *off screen*, to allow partial views of NATs at greater resolution (the % sign is optional).

```
        to reposition or rotate NATs enter 99, for default RETURN
```

If 99 is selected, the position and angle may be specified; for example,

```
        enter x position, (default 0%) displace from centre - or +:
        enter y position, (default 0%) displace from centre - or +:
```

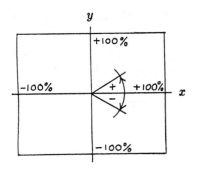

The default orientation of NATs may be rotated by a selected angle, + for anti-clockwise and − for clockwise. For *cycle lengths* greater than 2, the first transient tree (if any) is attached to the node lying due east (0 degrees). For cycles of 1 and 2, the default orientation is as illustrated in A1.3.

```
        alter angle of initial transient from cycle, for default RETURN:
        enter angle (degrees) anti-clockwise from +x axis:
```

A1.2.9 Alter Presentation

Various options are offered for altering the default presentation of NATs, which may be used in any combination (see A1.2.13 for examples). Suppressing copies of equivalent transient trees will result in drawing only the first of each equivalent set of transient trees (or branches from the uniform states, all 0s or 1s). However, garden-of-Eden nodes may be retained, as the footprints of the suppressed transient trees.

The option to mark singleton states (a single 1 among 0s) will put a white circle around such nodes. Numbering the nodes according to the global state will place the decimal equivalent of the binary state (or the binary state itself) at the node. Beyond a certain scale, about 12 levels (see A1.2.10), these numbers become indistinct.

Garden-of-Eden nodes may be reduced to a dot, or suppressed. Pre-image nodes (i.e., nodes in transient trees that are not garden-of-Eden nodes) may be suppressed. Increasing the angle between transient arcs can be useful for a clearer presentation, if pre-imaging is restricted (for example, Fig. 4.6).

```
        to alter presentation of NATs enter 99, else RETURN:
```

If 99 is selected, the following options are offered:

```
        to suppress equivalent transients enter 99, else RETURN:
        to mark singleton states enter 99, else RETURN:
        to number nodes according to state enter 99, else RETURN:
        to reduce size of Garden of Eden nodes enter 99,
              to suppress enter 88, else RETURN:
        to suppress pre-image nodes enter 99, else RETURN:
        to increase angle between pre-image arcs, enter 99, else RETURN:
```

A1.2.10 Set the Number of Levels (Scale)

This scales the NATs by setting the approximate number of levels (time steps to the attractor) that would fit the screen if a NAT was positioned centrally. The default varies according to previously selected parameters. For practical reasons of node density and transient length, the gap between levels decreases by a logarithmic function of the distance (in levels) from the attractor.

```
set number of levels, for default ( varies )-RETURN:
```

A1.2.11 Set the Separation between NATs

This may be required to amend the default separation between multiple NATs on the screen, for instance, for a basin of attraction field over a range of line lengths. An initial x- and y-axis separation factor is specified. A factor of 1 ensures that two NATs, each with transients trees about eight levels long, will not touch. As transients trees tend to become longer with increasing line length, the separation factor can be set to automatically increase by a given compound percentage.

```
enter x separation factor, for default (.5)-RETURN:
enter y separation factor, for default (.5)-RETURN:
```

If a range of line lengths was selected, specify the percentage increase (the % sign is optional)

```
enter compound increase x separation (default 10%):
enter compound increase y separation (default 10%):
```

A1.2.12 Select the Seed State for Single NAT(s)

Various options are offered for specifying the seed state for single NATs (NAT fields are seeded automatically). Any state that is part of the NAT will seed the entire NAT. The default is a singleton state. Note that specific binary or decimal seeds are selected on the graphics screen, in the top left dialogue area.

```
           select seed line
     1)    all 0s
     2)    all 1s
     3)    01.. repeating
     4)    001... repeating
     5)    0001.... repeating
     6)    singleton - positive (default)
     7)    singleton - negative
     8)    random
     9)    random-symmetric
     10)   decimal
     11)   binary
           enter 1-11 or RETURN:
```

If "transient branch only" was selected in A1.2.2, an option to start the transient branch from a seed at a given number of steps forward in time from the selected seed state is offered.

```
transient branch from future time step,
how many steps forward (default 0, max (varies)):
```

Running forward by at least one step before seeding the transient branch will ensure the existence of pre-images. A random state is very likely to have no pre-images (a garden-of-Eden state).

A1.2 ATLAS1: Basin of Attraction Fields

A1.2.13 The Graphics Screen and Examples

When the setup choices are complete, the following option to revise choices from A1.2.10, or RETURN for graphics, is presented:

```
99 to revise, RETURN for graphics:
```

After RETURN, the graphics screen will appear (black background), laid out as shown below. The attractor cycle will be generated first, followed by transient trees in turn, growing outward from the attractor. Successive NATs will be generated in successive rows from the left.

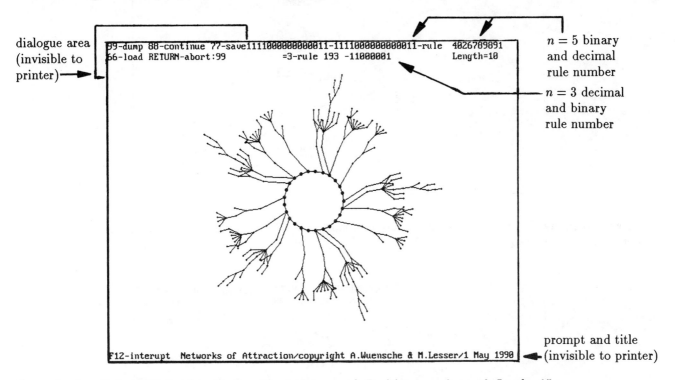

Example of a *single NAT*, line length, $L = 10$, singleton seed. Position: $x = 0$, $y = 0$. Levels: 15.

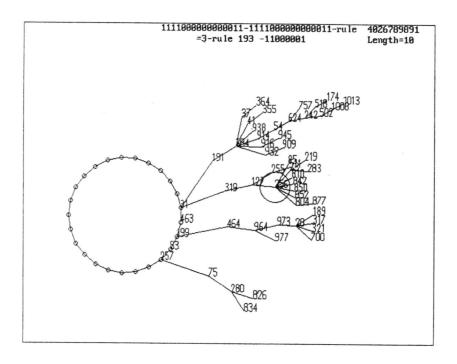

Example of *single NAT*, $L = 10$, transients suppressed, garden-of-Eden nodes suppressed, singleton marked, nodes numbered, singleton seed. Position: $z = -50$, $y = -20$. Angle: rotated -50 degrees. Levels: 4.

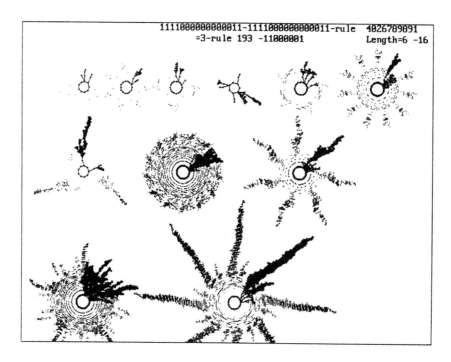

Example of single NATs, for range of $L = 6$ to 16, transients suppressed, garden-of-Eden nodes reduced, singleton seed. Position: $x = -70$, $y = 60$. Levels: 130. Separation: $x : 1$, $y : 1$. % increase: $x : 15$, $y : 15$.

A1.2 ATLAS1: Basin of Attraction Fields

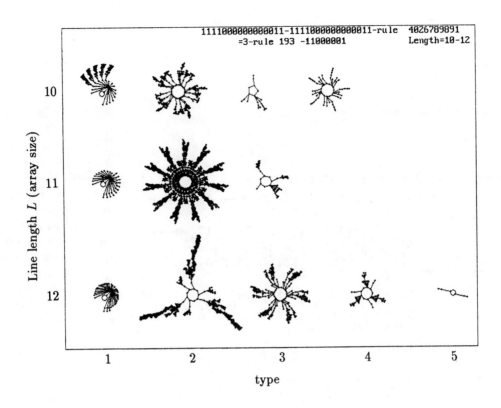

Example of *NAT fields*, for range of $L = 10$ to 12. Note that only one example of each set of *rotation equivalent* NATs is drawn. Printed data on these NAT fields (see A1.2.3), and key to this data, is shown in A1.2.14. Position: $x = -80$, $y = 60$. Levels: 120. Separation: $x : 1.8$, $y : 1.8$. % increase: $x : 8$, $y : 25$.

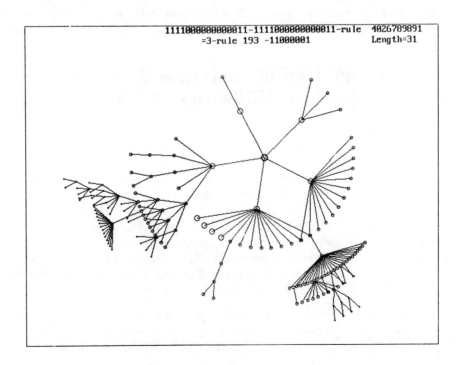

Example of a *transient branch only* from the seed two steps forward in time from the state 1997870096. $L = 31$. Position: $x = 15$, $y = 17$. Levels: 6.

A1.2.14 Key to Printed Data

If the "NAT info to printer" option (A1.2.3) is selected, data will be printed as NATs are drawn. The key to the printed data is set out below.

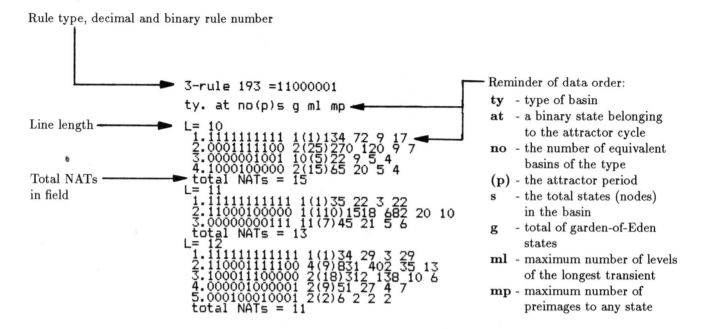

A1.2.15 The Interrupt Message: To Print, Load, or Save the Screen Image

Except when printing, loading, or saving, "key F12 + RETURN" will interrupt operation and the following interrupt message will appear in the top left-hand corner of the graphics screen, the dialogue area; in the setup screen the message appears in the text:

```
99-dump 88-continue 77-save
66-load RETURN-abort:99
```

RETURN-abort	▪ Returns to the beginning of the setup screen.
99-dump	▪ Prints the screen image to the printer. To continue from point of interrupt after printer dump, select 88 and RETURN twice. key F11 interrupts printing "error - 57 - printer not ready", may occur
88-continue	▪ Continues from the point of interrupt.
77-save	▪ Saves the screen image to disk.
66-load	▪ Loads a screen image file, or series of files, from disk. To continue from point of interrupt after loading, select 88 and RETURN twice.

The extension .NAT is automatically included in all graphics screen filenames. Omit the extension when entering a filename. Where indicated, *.NAT* files may be listed by responding with ? when prompted for the filename.

A1.2 ATLAS1: Basin of Attraction Fields
71

A1.2.16 Saving the Screen Image to Disk

at the interrupt message enter...	77
you will be prompted for...	`filename:`
answer with any legal name	
without extension, i.e....	d:rule193
while the file is being saved...	`d:rule193 - being saved`
will appear in the dialogue area.	

After saving, the interrupt message will reappear in the dialogue area. You may continue from the point of interrupt with 88.

A1.2.17 Automatic Screen Save

A sequence of graphics screens of a rule and its mutants may be saved automatically. This procedure is initiated in the setup screen (see A1.2.6).

Files will be saved automatically after each screen image is complete, with the mutant number added to the filename, i.e.,

myfile0	source rule
myfile1	1st mutant
myfile2	2nd mutant
⋮	⋮
myfile32	32nd mutant

If the start position was altered, for example, to 20, the file name of the first mutant will be "myfile20." A message in the dialogue area will indicate that a file is being saved, i.e., `d:myfile25-being saved`. When the last file has been saved, the interrupt message will reappear in the dialogue area.

A1.2.18 To Load a Screen Image File

at the interrupt message enter...	66
you will be prompted...	`for automutant-99 else RETURN:`
(for automutant see 2.19 below)	`filename (? to list, 99 to exit):`
(if the drive is omitted, the default is c:, or the last drive specified)	
enter the filename, i.e....	d:rule193
a chance for revision is offered...	`d:rule193.NAT -to revise enter 99:`
for overlay of images...	`keep screen-99 else RETURN:`

On RETURN, the image will be loaded. "`error - 53 - file not found`" may occur.

A1.2.19 To Load a Sequence of "Mutant" Files Automatically

at the interrupt message enter...	66
at the prompt...	`for automutant-99 else RETURN:`
enter 99...	`one screen-99 else RETURN:`
enter 99...	`filename (? to list, 99 to exit):`
enter the "mutant" filename,	
omitting the mutant number (see 2.17)...	d:myfile
a chance for revision is offered....	`d:myfile.NAT - to revise enter 99:`
you will be prompted...	`start mutant 1-32 (default=1):`
(select the start mutant number)	

72 APPENDIX 1 The Atlas Program

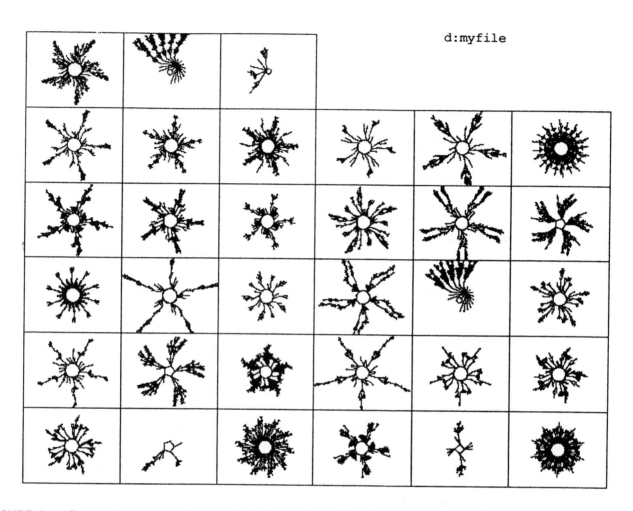

FIGURE A1.1 Composite screen: original rule, bottom left and the 32 mutants in ascending rows from left to right. Setup: rule 193, single NAT, $L = 10$, levels 15.

If "one screen" is selected, the sequence of mutant files will be loaded onto one composite screen, as shown in Fig. A1.1. Otherwise the "mutant" files will be loaded to the whole screen in sequence, with an optional pause between successive screens.

When the composite screen has been loaded, prompts appear in the top *right*-hand corner,

<pre>
to dump to the printer...   99 to dump:
to save the composite screen to disc... 99 to save:
if save is selected...   filename (default d:myfile-M):
enter any legal filename ending with -M, d:193-M
</pre>

A filename ending with -M will be recognised as a composite screen filename. To load a composite screen file, proceed as for loading a single screen image file (see A1.2.18).

A1.3 Graphic Conventions

A1.3.1 Point Attractors

A point attractor is represented as a node cycling to itself. The centreline of the pre-image fan (level 1) is set at a default angle of 45 degrees, and the fan is spread out as illustrated below, to minimise overlap to transient branches. The centre line of subsequent pre-image fans (level 2+) are radial to the point attractor node.

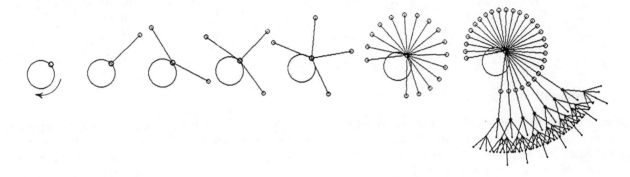

The pre-image fan (level 2) of a single pre-image (at level 1) of a point attractor is spread out on a circle centred at the point attractor node, and radial to the level 1 pre-image. The centre lines of subsequent pre-image fans (level 3) are radial to the point attractor node.

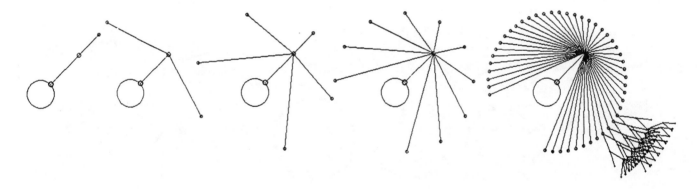

A1.3.2 Transient Branch Only

A NAT fragment representing an isolated transient branch from a given seed state is represented in a similar way to a point attractor, except that there is no indication of the seed node *cycling to itself*. The pre-image fan is evenly spaced around the seed node. (See the last example in A1.2.13.)

A1.3.3 Period-2 Attractors

A period-2 attractor has a default orientation and spread-out pre-image fans as illustrated below. pre-image fans at level 1 are angled relative to the attractor cycle to indicate that evolution proceeds clockwise around the attractor. The centre lines of subsequent pre-image fans (level 2+) are radial to the notional centre point of the attractor cycle.

74   APPENDIX 1 The Atlas Program

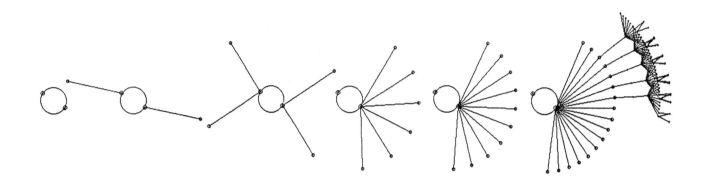

A1.3.4 Period-3+ Attractors

Attractor cycles with periods of 3 and above are represented as polygons. The diameter of the polygon asymptotically approaches an upper limit with increasing period, so that attractor cycles with larger periods are drawn with approximately the same diameter irrespective of the number of nodes in the attractor, as illustrated below.

Level 1 of each pre-image fan is angled relative to the centre of the polygon, to indicate that evolution proceeds clockwise around the cyclic attractor. By default, the first transient tree (if any) is attached to the node lying due east (0 degrees). The centre lines of subsequent pre-image fans (level 2+) are radial to the notional centre point of the attractor cycle.

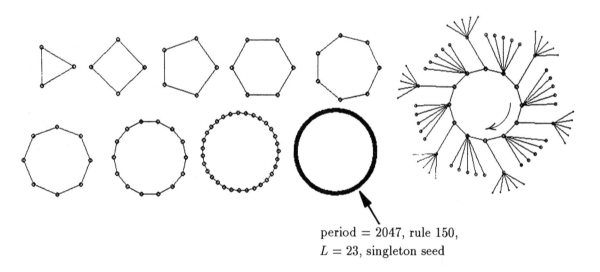

period = 2047, rule 150, $L = 23$, singleton seed

The default orientation of NATs may be rotated (see A1.2.9).

In transient trees, the gap between successive levels (representing time steps) decreases by a logarithmic function of the distance (in levels) from the attractor.

A1.3.5 Colours

Four colours—red, blue, green and yellow—are used for lines representing transitions (and four contrasting colours for nodes). Equivalent transients trees (and equivalent branches from uniform states) are coloured identically.

For attractor periods of 6 and above, successive nonequivalent transients trees (and branches) will be assigned the next available colour, thus cycling through the four colours. For attractor periods of 5 and below, successive fans will be assigned the next available colour.

Garden-of-Eden nodes and attractor cycle nodes without transients are coloured white.

A1.4 SPACE1: Space-Time Patterns

A1.4.1 Running the Program; the Set-up Screen

To run: insert the floopy disk into a drive, or save onto hard disk; if in drive A, at the DOS prompt, enter A:SPACE1. The setup screen (grey background) will appear, with this message:

```
free date bytes 503696 (or similar), at least 180000 required.
to quite Space1 enter 99 + RETURN, else RETURN:
```

The setup choices, as they would appear on screen, are summarised below, with a brief commentary on each choice.

In general, 99 selects an option, and RETURN the default. A series of prompts are presented to specify the pattern mode, rule parameters, and the seed line.

A1.4.2 Select Space-Time Pattern Mode

```
1. text mode (fast)..................max L -  80
2. graphics mode.....................max L -  79
3. graphics pixel mode (default)......max L - 640
enter 1, 2 or 3:
```

MODES 1 AND 2. The operation of modes 1 and 2 are basically the same, with space-time patterns presented as text on a scrolling screen. The differences between the two modes are as follows:

- Mode 1 is much faster than mode 2 (too fast for some applications).
- Mode 1 has 25 lines of scrolling text, whereas mode 2 has 29 lines.
- Mode 2 can be dumped to printer; mode 1 cannot.

RUNNING "FORWARDS" OR "BACKWARDS." In modes 1 and 2 (running "forward"), a repeat state, and thus the attractor cycle, may be identified. The CA may also be run "backwards"; pre-images (and pre-images of pre-images) may be generated from any seed state, thus potentially disclosing the seed's entire upstream configuration. If the seed state is on an attractor cycle, this configuration is the basin of attraction. If the seed is on a transient tree, the configuration is the transient branch joining the seed.

When running "forward" in modes 1 and 2, the decimal equivalent of the CA's global state will be given alongside the space-time pattern, up to $L = 53$.

MODE 3. In mode 3, space-time patterns are presented in high resolution on the basis of single screen pixels. The line length may be set to a maximum of 640. Patterns do not scroll; after 480 time steps, which fill the screen vertically, evolution of the CA is continued from the top of the screen by pressing RETURN.

Mode 3 does not allow for identifying repeat states, or running "backwards."

A1.4.3 Select the Type of Rule and Notation

The rule type may be selected as an $n = 5$ rule (in decimal, or binary), an $n = 3$ rule, or an $n = 5$ totalistic code (the specific rule will be chosen in A1.4.4). The default is an $n = 5$ decimal rule.

```
Select notation to specify 5-rule or subset
1) 5-rule Decimal(default)
2) 5-rule Binary
3) 3-rule
4) 5-rule Totalistic code
enter 1, 2, 3, 4, 5 or RETURN:
```

A1.4.4 Select the Rule

According to the choice of rule type and notation in section A1.4.3, the rule is selected in decimal or binary (the default is a random rule). For example,

```
select 3-rule notation 1) decimal (default), 2) binary
enter 1, 2 or RETURN:

enter 3-rule number 0 to 255 (default random):
enter 0s and 1s to make up an 8-bit binary number:
```

The rule table, the λ parameter, the λ ratio, and the Z parameter will be given for the chosen rule.

A1.4.5 Select the Line Length

The minimum line length is 1; the maximum line length (and default) depends on the mode:

- mode 1 - max L and default = 80
- mode 2 - max L and default = 79
- mode 3 - max L = 640, default = 150

For example,

```
select first line length (max 640) default 150:
```

In modes 1 and 2 only, the number of time steps for which the repeat check is active may be set. This allows attractor cycles to be identified. The default is 200. If 0 is set, the repeat check is deactivated.

```
extent of repeat check (default 200):
```

A1.4.6 Mutation and Automatic Mutation

While the CA space-time pattern is being generated on screen, a one-bit mutation may be made at a random position in the $n = 5$ rule table by the pressing *key F1*. The rule just changed is restored by pressing *key F2*. This may be done repeatedly to select interesting patterns. At each mutation, the colour of the space-time pattern changes from yellow to blue, or vice versa. In *mode 3, complex rules* may be easily found using this method.

If *automatic mutation* is selected, the rule table will randomly mutate by one bit at every time step. During the run, *key F1* will toggle automatic mutation on and off. This will also change the colour of the space-time pattern between yellow and blue.

```
set automatic mutation enter 99:
```

To summarise, by default,   **key F1** - one-bit random mutation
                                 key F2 - restores rule just changed
If automatic mutation is set,   **key F1** - toggles automatic mutation on and off

A1.4.7 Select the Seed State

Various options are offered for specifying the seed state. The default is random. In mode 3, the binary option allows a central zone of the seed line (max 79 cells) to be defined, with other cells set to 0.

A1.4 SPACE1: Space-Time Patterns

```
enter 0s and 1s to make up a line  78  characters long
------------------------------------------------------------------------------
100000000000010000000000001000000000000100000000000010000000000001000000000000
8............8............8............8............8............8...........
111111111111100100100100101001001001001111111111111100100100100101001001001001
010010010010011111111111111001001001001010010010010011111111111111100100100100
001001001001000100100100100010010010010001001001001000100100100100010010010010
100100100100101001001001001111111111111100100100100101001001001001111111111111
alternative 1
⬛⬛⬛⬛⬛⬛⬛⬛⬛⬛⬛⬛⬛..8..8..8..8.8..8..8..8.⬛⬛⬛⬛⬛⬛⬛⬛⬛⬛⬛⬛⬛..8..8..8..8.8..8..8..8..
110110110110100111100111100010100010100000000000000010100010100011110011110011
011011011011011100111100110011110011110110110110110100010100010101000101000101
000000000000000101000101000111100111100110110110110100111100111100010100001010
101101101101010100010100010100010110110110110110111001110011100111100111100111
alternative 2
.8..8..8..8..⬛⬛⬛⬛⬛⬛⬛⬛⬛⬛⬛⬛⬛..8..8..8..8.8..8..8..8..⬛⬛⬛⬛⬛⬛⬛⬛⬛⬛⬛⬛⬛.8..8..8..8
011110011110011011011011011001111001111000101000101000000000000000101000101001
110011110011110110110110110100010100010101000101000101101101101101111001111001
101000101000101101101101101111001111001110011110011110110110110110110001010001
000101000101000000000000000101000101000111100111100110110110110110011110011110
alternative 3
```

FIGURE A1.2 Mode 2: "backwards." Example of running "backwards" from a given seed line, length 78, $n = 3$ rule 105. Mode 1 is similar but faster, and may not be printed.

```
select seed line
1) Singleton
2) Random (default)
3) Random-symmetric
4) Decimal
5) Binary
enter 1, 2, 3, 4 or 5:
```

A1.4.8 Pre-images

As described in A1.4.2, in modes 1 and 2 only, the pre-images (and pre-images of pre-images) will be generated if running "backwards" is selected:

```
Run CA forwards(default)-1 or backwards-2, enter 1 or 2:
```

If "backwards" is selected, a further option is offered, to start running backwards from a seed at a given number of steps forward in time from the selected seed state.

```
forward before running backward,
how many steps forward (default 0, max (varies)):
```

Running forward by at least one step before running backwards will ensure the existence of pre-images. A random state is very likely to have no pre-images (a garden-of-Eden state).

When run "backwards," "error - 14 - out of memory" may occur, if the configuration of rule and seed, combined with an excessive line length, results in too many *partial pre-images* or *pre-images*.

FIGURE A1.3 Mode 2: "forwards." Example of running "forwards" from a given seed line, length 78, $n = 3$ rule 105. Mode 1 is similar but faster, and may not be printed.

A1.4.9 The Graphics Screen and Examples

When the setup options have been selected, the space-time pattern will be generated on a black background. Modes 1 and 2 will scroll the space-time pattern; mode 3 will present successive screens. See Figures A1.2–5.

A1.4.10 The Interrupt Message, the Revision Menu, and Printing

Except when printing, **key F12 + RETURN** will interrupt operation and the interrupt message will appear in the bottom right-hand corner of the graphics screen, the dialogue area; in the setup screen the message appears in the text:

continue-RETURN	▪ Continues from the point of interrupt.
printer dump-77 (not in mode 1)	▪ Prints the screen image to the printer. The latest (mutated) rule number (and λ, λ ratio and Z) will be given. To continue after printer dump, select 88 then RETURN twice. **Key F11** interrupts printing "error - 57 - printer not ready", may occur
revision menu-88	▪ See A1.4.11 below
start menu-99	▪ Returns to the beginning of the setup screen.

There is no facility to save and load screen images in Space1. The graphics are typically generated as quickly in the program as loading a file.

A1.4.11 The Revision Menu

The revision menu is presented after being selected from the interrupt options, or, if a repeat state has been identified, in modes 1 and 2. The revision menu allow a new seed state and new line length to be selected, for the current rule.

Color Plates

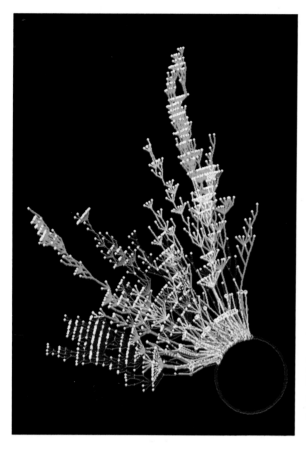

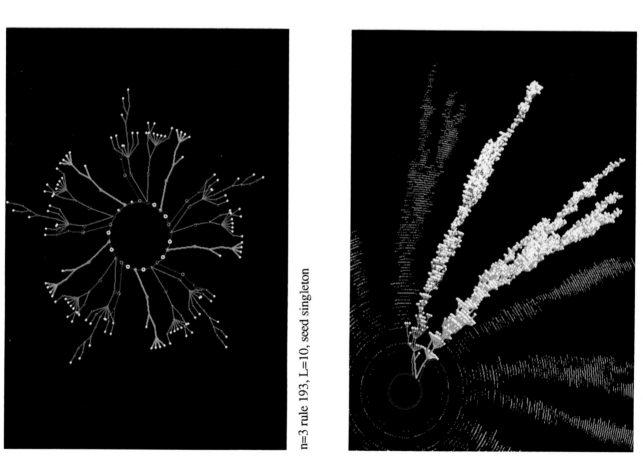

n=3 rule 193, L=10, seed singleton

n=3 rule 193, L=18, seed 011010001110000010
Equivalent transient trees suppressed.

n=3 rule 193, L=15, seed 110011000001101
Equivalent transient trees suppressed.

n=3 rule 41 (mutant 1), L=16, seed 0001010001000101

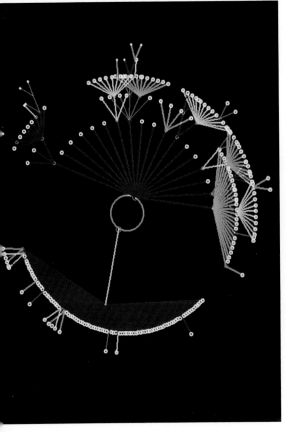

n=3 rule 33, L=16, seed 0111111111100000

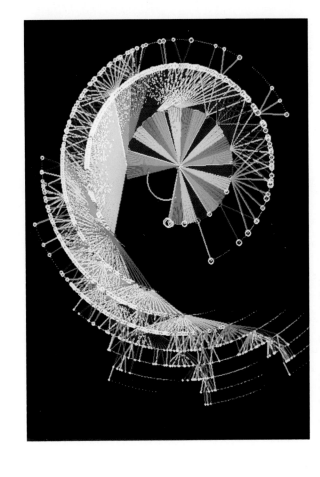

n=3 rule 249, L=15, seed all 0s
Equivalent transient branches suppressed.

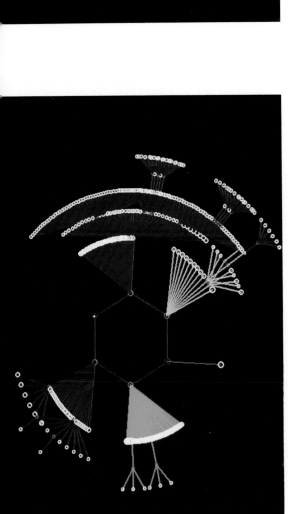

n=3 rule 18, L=18, seed 101000001010000000

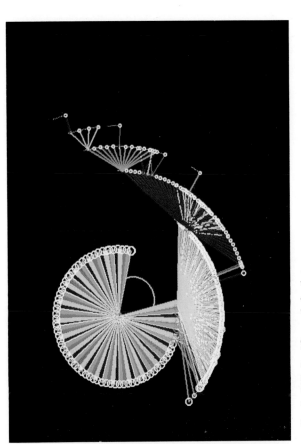

n=3 rule 251, L=12, seed singleton
Equivalent transient branches suppressed.

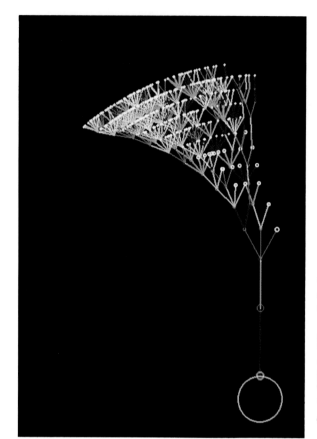

n=3 rule 250, L=15, seed all 1s
Equivalent transient branches suppressed.

n=3 rule 228, L=15, seed singleton

n=3 rule 9, L=12, seed 010000010001

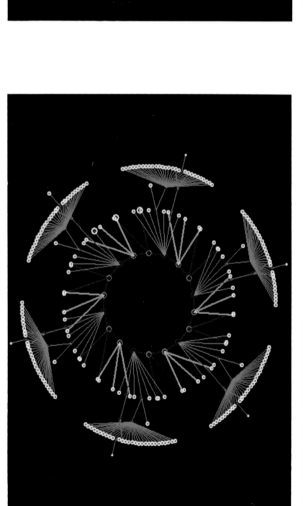

n=3 rule 58, L=15, seed 011011011011011

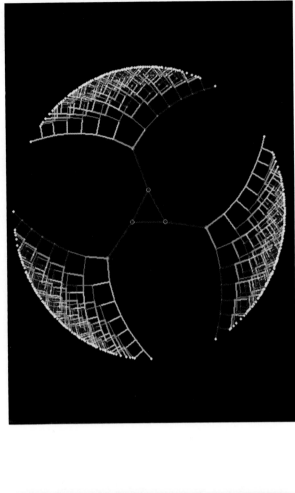

n=3 rule 60, L=24, seed 011011011011011011011011

n=3 rule 51 (mutant 1), L=31, seed singleton

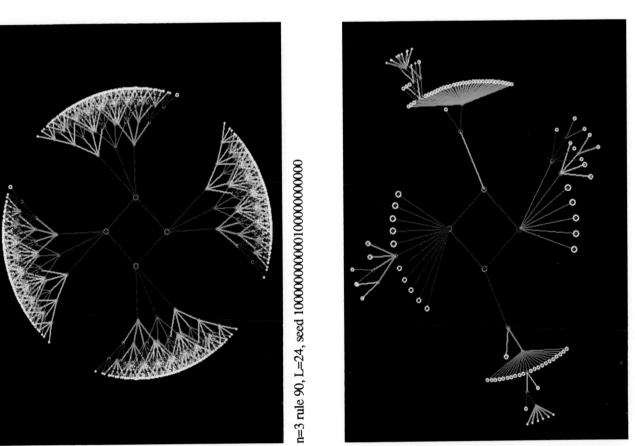

n=3 rule 90, L=24, seed 100000000000100000000000

n=3 rule 126, L=11, seed singleton

n=3 rule 225, L=13, seed 2769

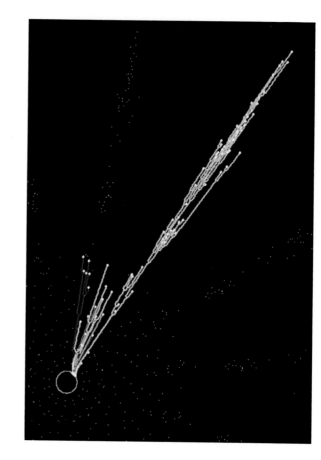

n=3 rule 30, L=15, seed 28096
Equivalent transient trees suppressed.

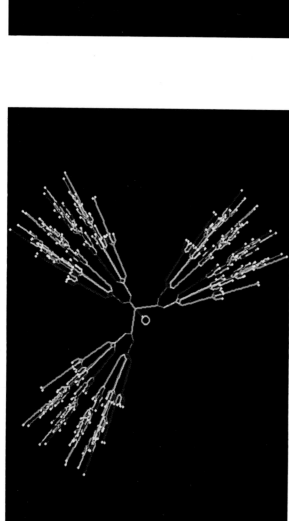

n=3 rule 225, L=12, seed all 0s

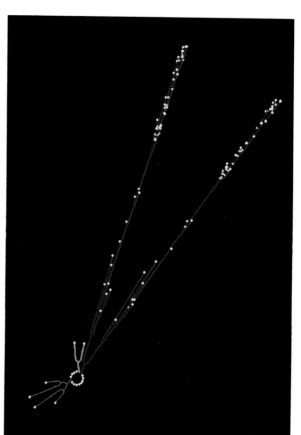

n=3 rule 225, L=16, seed 1000000010000000

n=5 code 10, L=15, seed singleton

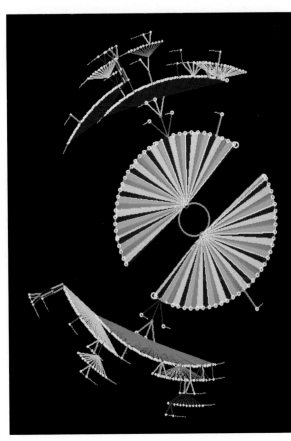

n=5 code 25, L=15, seed all 0s
Equivalent transient branches suppressed.

n=5 code 10, L=16, seed singleton

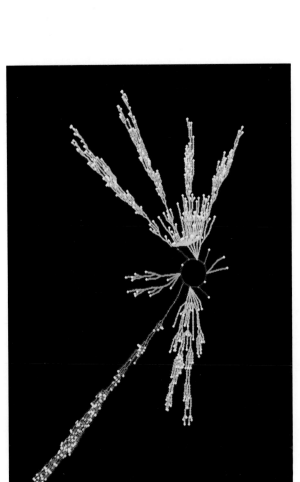

n=5 code 10, L=19, seed singleton

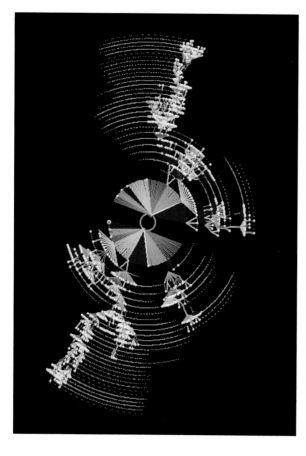

n=5 code 11, L=15, seed all 0s
Equivalent transient branches suppressed.

n=5 code 53, L=15, seed singleton

n=5 code 14, L=12, seed all 0s
Equivalent transient branches suppressed.

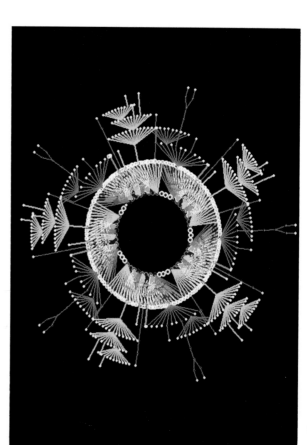

n=5 code 53, L=15, seed 101111101101100

A1.4 SPACE1: Space-Time Patterns

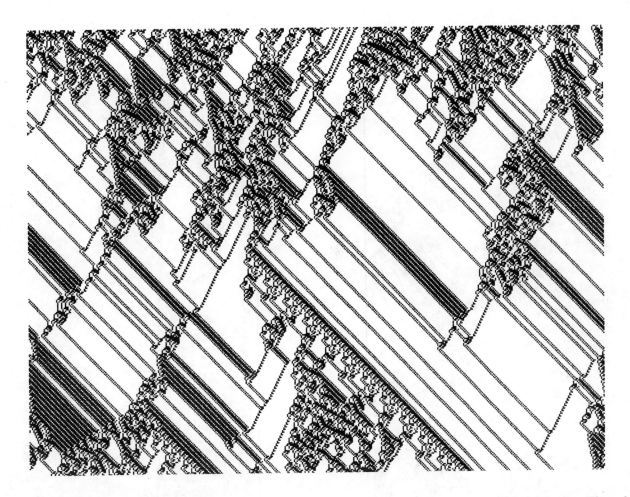

FIGURE A1.4 Mode 3. $n = 5$ rule 4234825112 ($Z = .6171875$, λ ratio $= .9375$), $L = 450$, 330 time steps. The rule "evolved" as illustrated in Fig. A1.5(a).

The latest (mutated) rule number (in decimal and binary), and the λ parameter, λ ratio, and the Z parameter will be given.

When running forward using *modes 1 and 2 only*, if a repeat state is found, the space-time patterns will stop. Information regarding the attractor cycle will be given.

The revision menu will then be presented, with the added options of selecting the *repeat state* as the new seed, and running forwards or backwards. A repeat state must be on the attractor cycle and thus have at least one pre-image.

A1.4.12 Color Cells According to Neighbourhoods

Each cell may be assigned a color depending on the neighbourhood which determined the cell's value. Colors are assigned to the rule table from a palette of 16 available colors. Black represents the neighbourhood all zeros, and white all ones. The color option is selected in the set-up screen, or toggled on and off during the run with key F3.

(a)

(b)

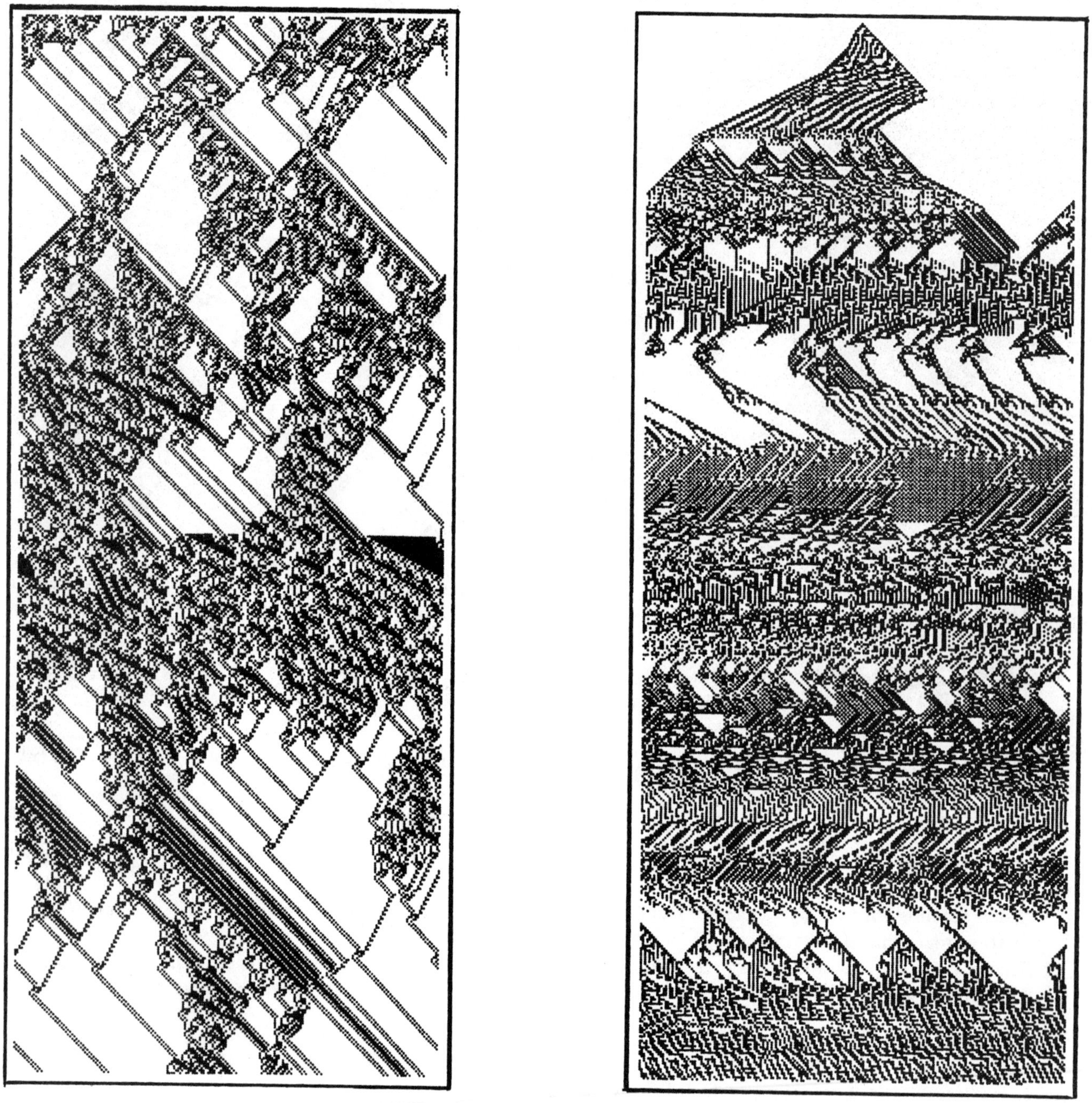

FIGURE A1.5 (a) Mode 3: one-bit mutations. 5-rule 1550470552 (see chapter 4, Fig. 4.1(b)) with a number of one-bit mutations during the run to "evolve" the $n = 5$ rule 4234825112, (see Fig. A1.4). 480 time steps. (b) Mode 3: automatic mutations. $n = 3$ rule 30, with a one-bit random mutation at each time step. $L = 200$. 480 time steps.

APPENDIX 2
Atlas of Basin of Attraction Fields

CA rules are presented in sequence according to the rule in the top left hand corner of the rule cluster. Complementary equivalence classes are presented on facing pages. Subject to available space, basin of attraction fields and data are shown for a range of array length L of 1 to 15 for $n = 3$ rules, and 3 to 16 for $n = 5$ totalistic codes. All basin of attraction fields are drawn to the same scale. The scale of the blow up of a typical basin varies.

Note that some pages are left intentionally blank in order for rules and their complements to appear on facing pages.

A2.1 Contents

A2.2 Key to Basin Field Presentation 82

A2.3 **n=3 Rules**
    A2.3.1 Index 83
    A2.3.2 Symmetrical Rules 84
    A2.3.3 Semi-Asymmetrical Rules 125
    A2.3.4 Fully Asymmetrical Rules 158

A2.4 **n=5 Rules, Totalistic Code**
    A2.4.1 Index 185
    A2.4.2 Totalistic Codes 185

82 APPENDIX 2 Atlas of Basin of Attraction Fields

A2.2 Key to Basin Field Presentation

A typical page from the Atlas of basin of attraction fields is annotated below.

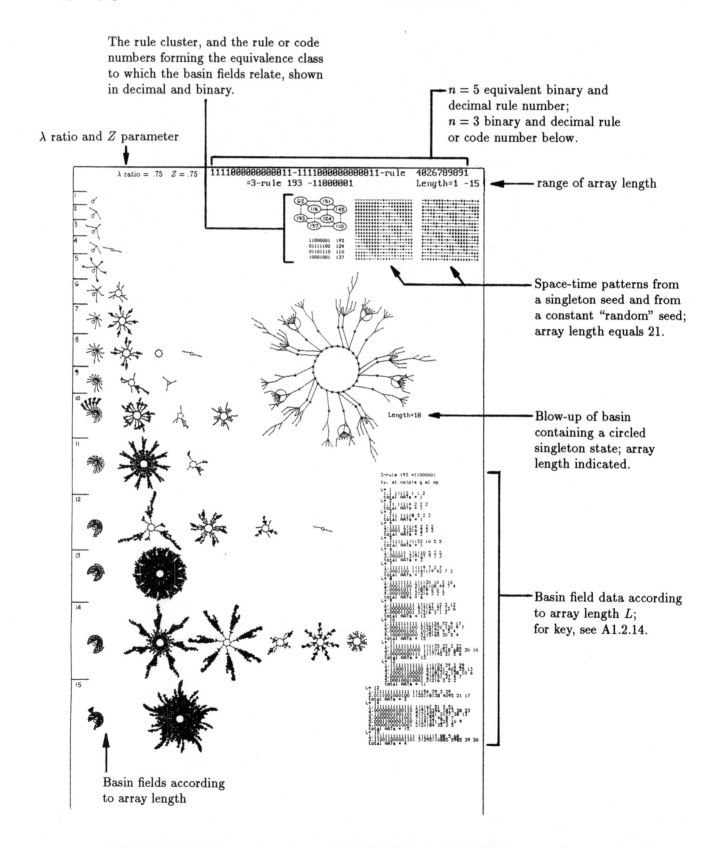

A2.3 n=3 Rules

A2.3.1 Index

Key: [decimal rule number],*[hex rule number]*-[page number].

0,*0*-85	32,*20*-89	64,*40*-127	96,*60*-131	128,*80*-87	160,*A0*-91	192,*C0*-129	224,*E0*-133
1,*1*-86	33,*21*-100	65,*41*-134	97,*61*-150	129,*81*-123	161,*A1*-119	193,*C1*-157	225,*E1*-145
2,*2*-126	34,*22*-137	66,*42*-169	98,*62*-173	130,*82*-135	162,*A2*-139	194,*C2*-171	226,*E2*-175
3,*3*-128	35,*23*-146	67,*43*-170	99,*63*-180	131,*83*-156	163,*A3*-154	195,*C3*-183	227,*E3*-173
4,*4*-88	36,*24*-103	68,*44*-136	100,*64*-149	132,*84*-101	164,*A4*-105	196,*C4*-147	228,*E4*-143
5,*5*-90	37,*25*-104	69,*45*-138	101,*65*-152	133,*85*-118	165,*A5*-117	197,*C5*-155	229,*E5*-141
6,*6*-130	38,*26*-148	70,*46*-172	102,*66*-183	134,*86*-151	166,*A6*-153	198,*C6*-181	230,*E6*-171
7,*7*-132	39,*27*-142	71,*47*-174	103,*67*-170	135,*87*-144	167,*A7*-140	199,*C7*-172	231,*E7*-169
8,*8*-127	40,*28*-131	72,*48*-93	104,*68*-97	136,*88*-129	168,*A8*-133	200,*C8*-95	232,*E8*-99
9,*9*-134	41,*29*-150	73,*49*-112	105,*69*-120	137,*89*-157	169,*A9*-145	201,*C9*-111	233,*E9*-97
10,*A*-161	42,*2A*-165	74,*4A*-141	106,*6A*-145	138,*8A*-163	170,*AA*-167	202,*CA*-143	234,*EA*-133
11,*B*-162	43,*2B*-176	75,*4B*-152	107,*6B*-150	139,*8B*-179	171,*AB*-165	203,*CB*-149	235,*EB*-131
12,*C*-136	44,*2C*-149	76,*4C*-107	108,*6C*-111	140,*8C*-147	172,*AC*-143	204,*CC*-109	236,*EC*-95
13,*D*-138	45,*2D*-152	77,*4D*-114	109,*6D*-112	141,*8D*-155	173,*AD*-141	205,*CD*-107	237,*ED*-93
14,*E*-164	46,*2E*-179	78,*4E*-155	110,*6E*-157	142,*8E*-177	174,*AE*-163	206,*CE*-147	238,*EE*-129
15,*F*-166	47,*2F*-162	79,*4F*-138	111,*6F*-134	143,*8F*-164	175,*AF*-161	207,*CF*-136	239,*EF*-127
16,*10*-126	48,*30*-137	80,*50*-161	112,*70*-165	144,*90*-135	176,*B0*-139	208,*D0*-163	240,*F0*-167
17,*11*-128	49,*31*-146	81,*51*-162	113,*71*-176	145,*91*-156	177,*B1*-154	209,*D1*-179	241,*F1*-165
18,*12*-92	50,*32*-106	82,*52*-140	114,*72*-154	146,*92*-113	178,*B2*-115	210,*D2*-153	242,*F2*-139
19,*13*-94	51,*33*-108	83,*53*-142	115,*73*-146	147,*93*-110	179,*B3*-106	211,*D3*-148	243,*F3*-137
20,*14*-130	52,*34*-148	84,*54*-164	116,*74*-179	148,*94*-151	180,*B4*-153	212,*D4*-177	244,*F4*-163
21,*15*-132	53,*35*-142	85,*55*-166	117,*75*-162	149,*95*-144	181,*B5*-140	213,*D5*-164	245,*F5*-161
22,*16*-96	54,*36*-110	86,*56*-144	118,*76*-156	150,*96*-121	182,*B6*-113	214,*D6*-151	246,*F6*-135
23,*17*-98	55,*37*-94	87,*57*-132	119,*77*-128	151,*97*-96	183,*B7*-92	215,*D7*-130	247,*F7*-126
24,*18*-169	56,*38*-173	88,*58*-141	120,*78*-145	152,*98*-171	184,*B8*-175	216,*D8*-143	248,*F8*-133
25,*19*-170	57,*39*-180	89,*59*-152	121,*79*-150	153,*99*-183	185,*B9*-173	217,*D9*-149	249,*F9*-131
26,*1A*-140	58,*3A*-154	90,*5A*-117	122,*7A*-119	154,*9A*-153	186,*BA*-139	218,*DA*-105	250,*FA*-91
27,*1B*-142	59,*3B*-146	91,*5B*-104	123,*7B*-100	155,*9B*-148	187,*BB*-137	219,*DB*-103	251,*FB*-89
28,*1C*-172	60,*3C*-183	92,*5C*-155	124,*7C*-157	156,*9C*-181	188,*BC*-171	220,*DC*-147	252,*FC*-129
29,*1D*-174	61,*3D*-170	93,*5D*-138	125,*7D*-134	157,*9D*-172	189,*BD*-169	221,*DD*-136	253,*FD*-127
30,*1E*-144	62,*3E*-156	94,*5E*-118	126,*7E*-123	158,*9E*-151	190,*BE*-135	222,*DE*-101	254,*FE*-87
31,*1F*-132	63,*3F*-128	95,*5F*-90	127,*7F*-86	159,*9F*-130	191,*BF*-126	223,*DF*-88	255,*FF*-85

A2.3.2 Symmetric Rule Clusters (see section 3.3.8)

By definition, $R = R_r$, so the reflection links will collapse.

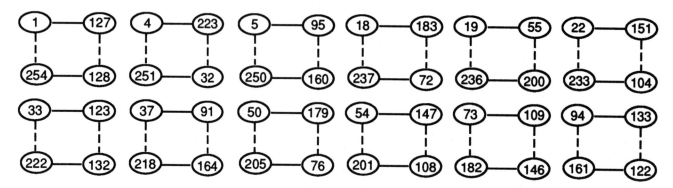

The cluster will collapse further, if for a given rule R, $R_c = R_n$

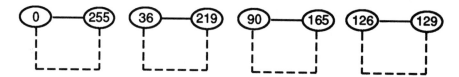

and also if $R = R_n$

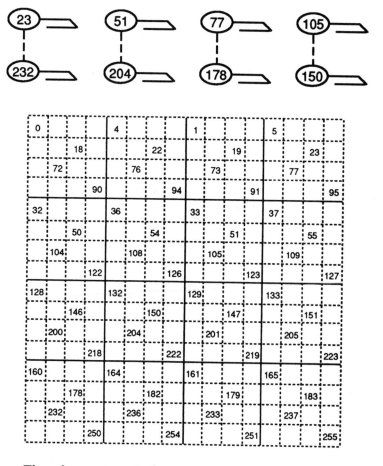

The rule-space matrix (see appendix 4).

85

λ ratio = 0 $Z = 0$ 0000000000000000-0000000000000000-rule 0
=3-rule 0 -00000000   Length=1 -8

00000000 0
11111111 255

Length=8

3-rule 0 =00000000
ty. at no(p)s g ml mp
L= 1
 1.0 1(1)2 1 1 2
 total NATs = 1
L= 2
 1.00 1(1)4 3 1 4
 total NATs = 1
L= 3
 1.000 1(1)8 7 1 8
 total NATs = 1
L= 4
 1.0000 1(1)16 15 1 16
 total NATs = 1
L= 5
 1.00000 1(1)32 31 1 32
 total NATs = 1
L= 6
 1.000000 1(1)64 63 1 64
 total NATs = 1
L= 7
 1.0000000 1(1)128 127 1 128
 total NATs = 1
L= 8
 1.00000000 1(1)256 255 1 256
 total NATs = 1

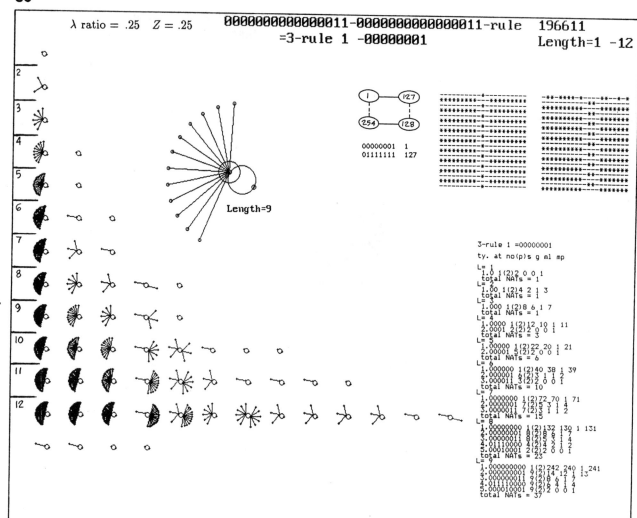

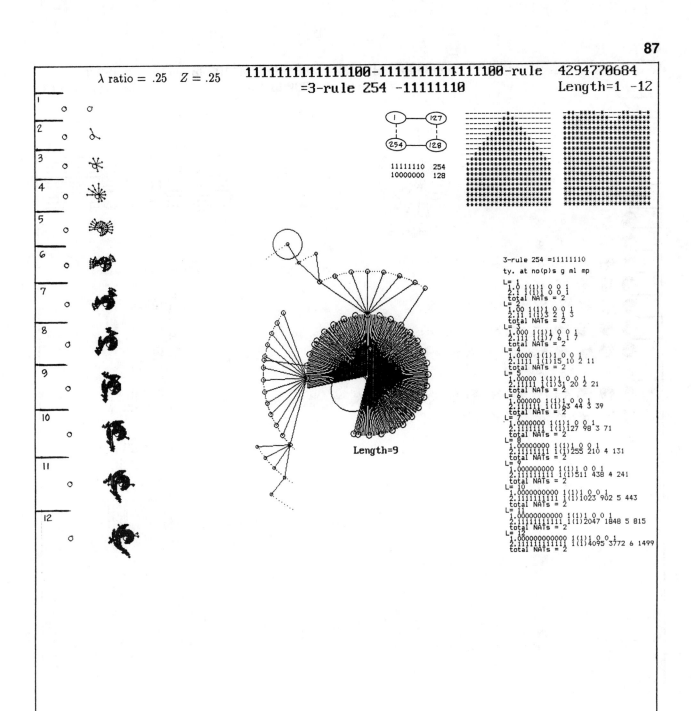

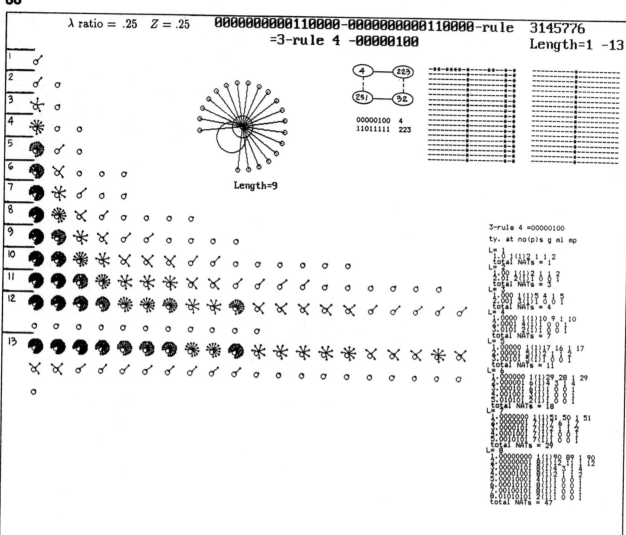

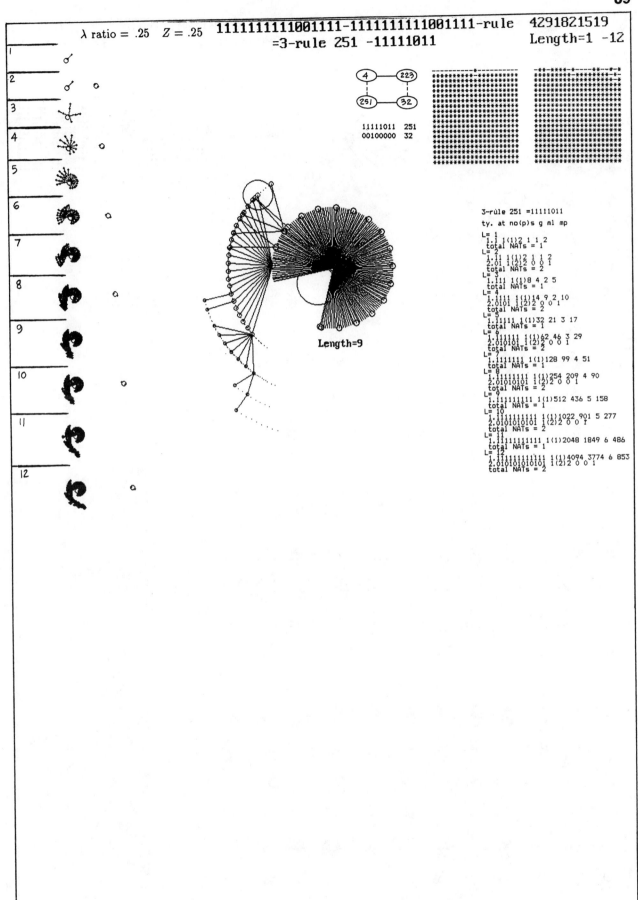

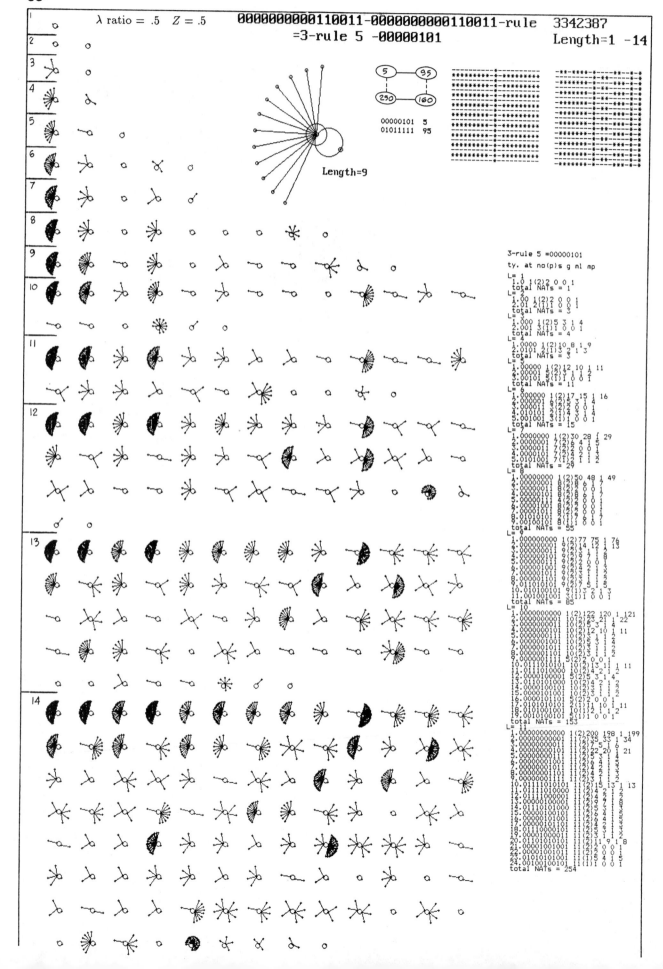

λ ratio = .5 Z = .5   1111111111001100-1111111111001100-rule   4291624908
=3-rule 250 -11111010   Length=1 -15

11111010 250
10100000 160

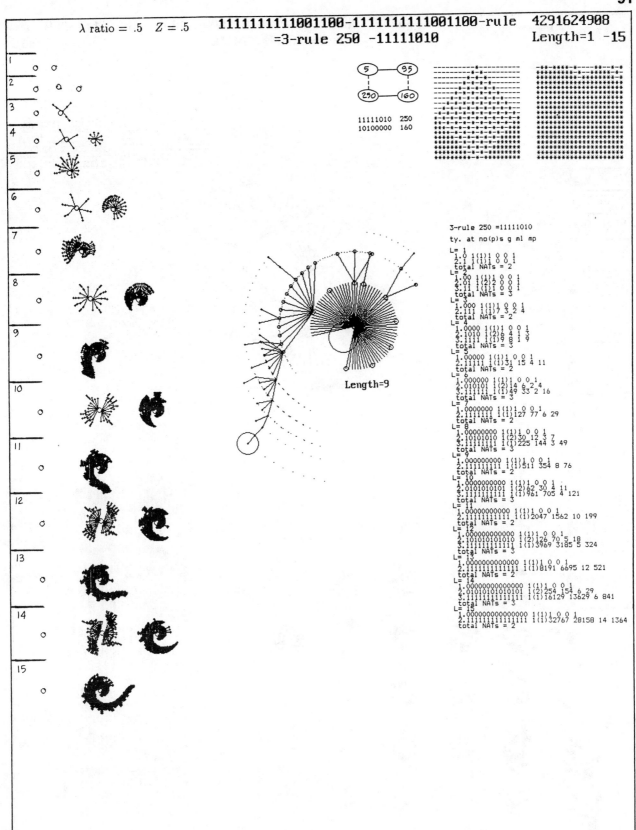

Length=9

3-rule 250 =11111010
ty. at no(p)s g ml mp
L= 1
1.0   1(1)1 0 0 1
2.1   1(1)1 0 0 1
total NATs = 2
L= 2
1.00 1(1)1 0 0 1
2.01 1(2)2 0 0 1
3.11 1(1)1 0 0 1
total NATs = 3
L= 3
1.000 1(1)1 0 0 1
2.111 1(1)7 3 2 4
total NATs = 2
L= 4
1.0000 1(1)1 0 0 1
2.1010 1(2)6 4 1 3
3.1111 1(1)19 8 1 9
total NATs = 3
L= 5
1.00000 1(1)1 0 0 1
2.11111 1(1)31 15 4 11
total NATs = 2
L= 6
1.000000 1(1)1 0 0 1
2.010101 1(2)14 6 2 4
3.111111 1(1)49 33 2 16
total NATs = 3
L= 7
1.0000000 1(1)1 0 0 1
2.1111111 1(1)127 77 6 29
total NATs = 2
L= 8
1.00000000 1(1)1 0 0 1
2.10101010 1(2)30 12 3 7
3.11111111 1(1)225 144 3 49
total NATs = 3
L= 9
1.000000000 1(1)1 0 0 1
2.111111111 1(1)511 354 8 76
total NATs = 2
L= 10
1.0000000000 1(1)1 0 0 1
2.0101010101 1(2)82 30 4 11
3.1111111111 1(1)961 705 4 121
total NATs = 3
L= 11
1.00000000000 1(1)1 0 0 1
2.11111111111 1(1)2047 1562 10 199
total NATs = 2
L= 12
1.000000000000 1(1)1 0 0 1
2.101010101010 1(2)126 70 5 18
3.111111111111 1(1)3969 3185 5 324
total NATs = 3
L= 13
1.0000000000000 1(1)1 0 0 1
2.1111111111111 1(1)8191 6695 12 521
total NATs = 2
L= 14
1.00000000000000 1(1)1 0 0 1
2.01010101010101 1(2)254 154 6 29
3.11111111111111 1(1)16129 13629 6 841
total NATs = 3
L= 15
1.000000000000000 1(1)1 0 0 1
2.111111111111111 1(1)32767 28158 14 1364
total NATs = 2

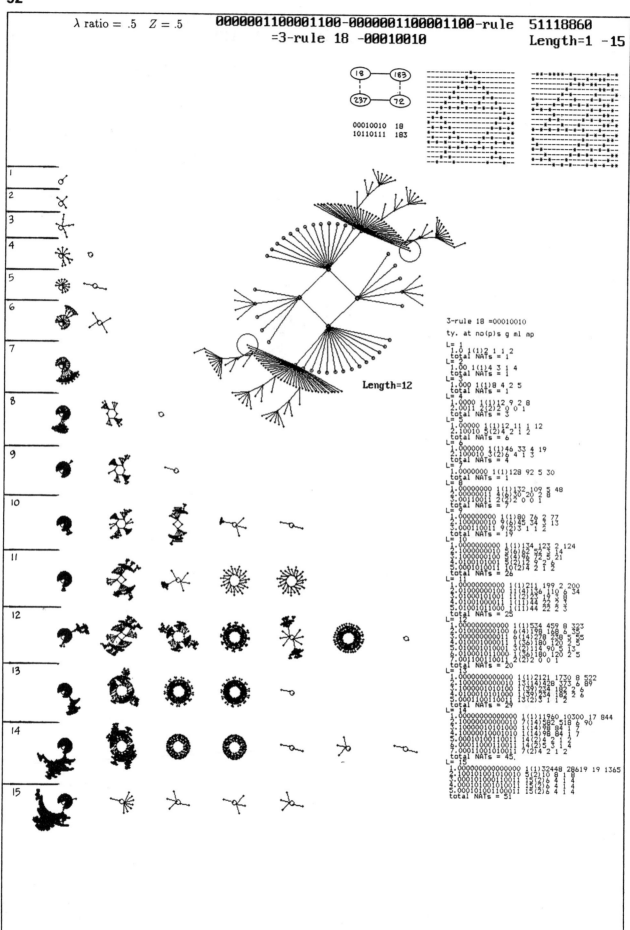

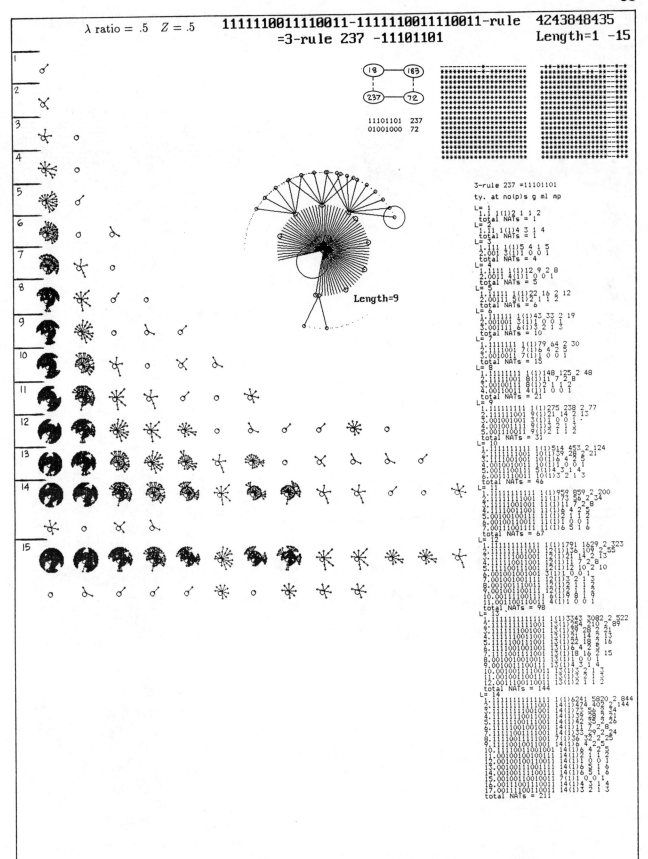

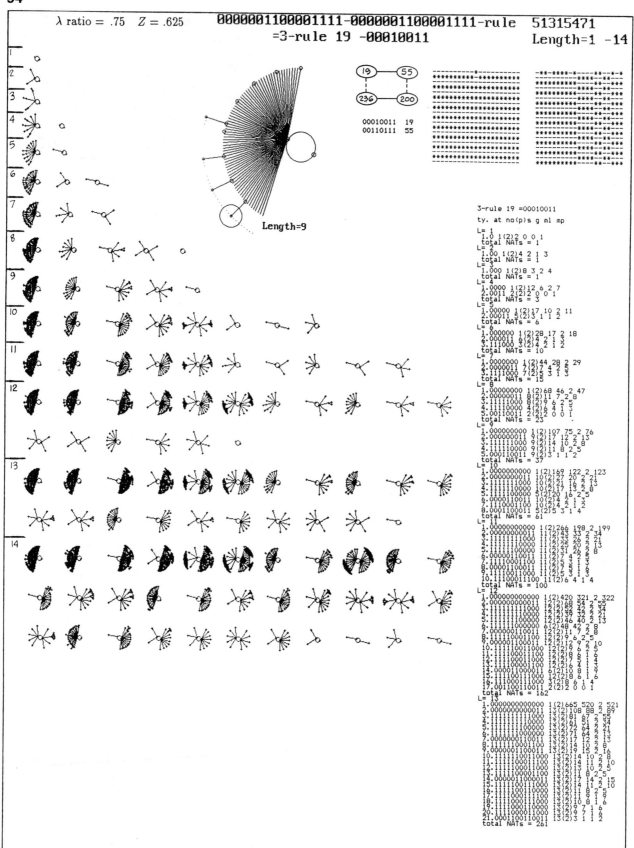

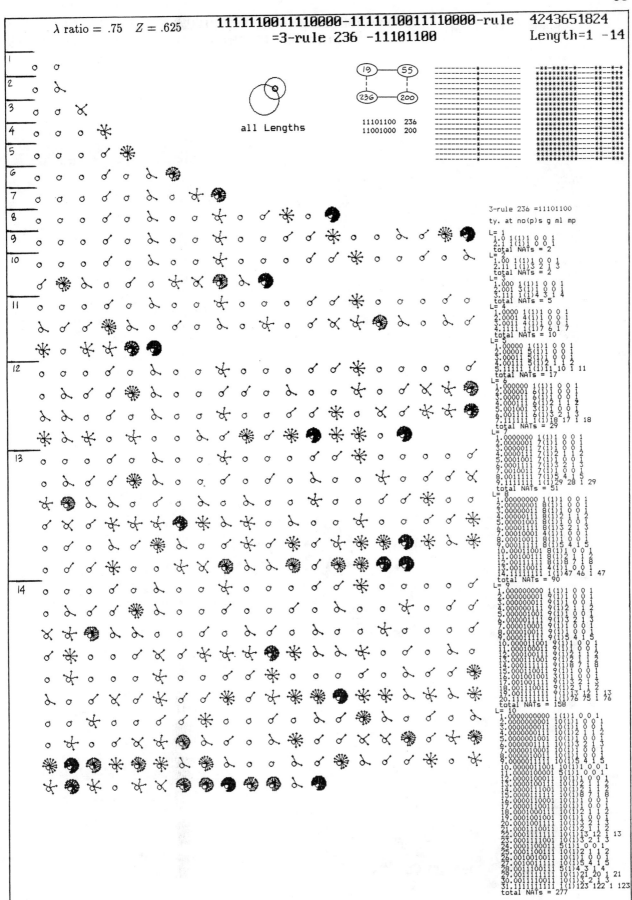

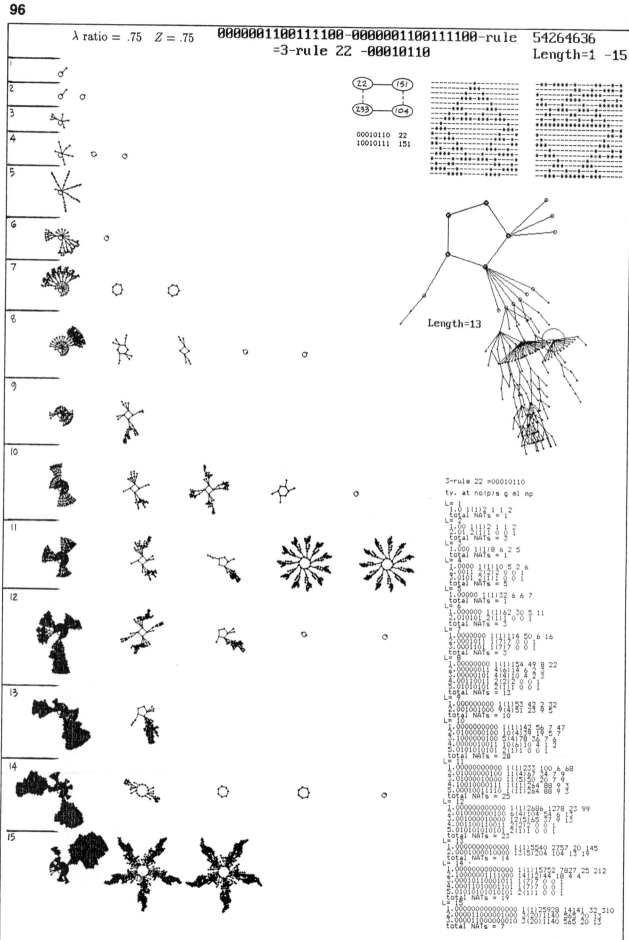

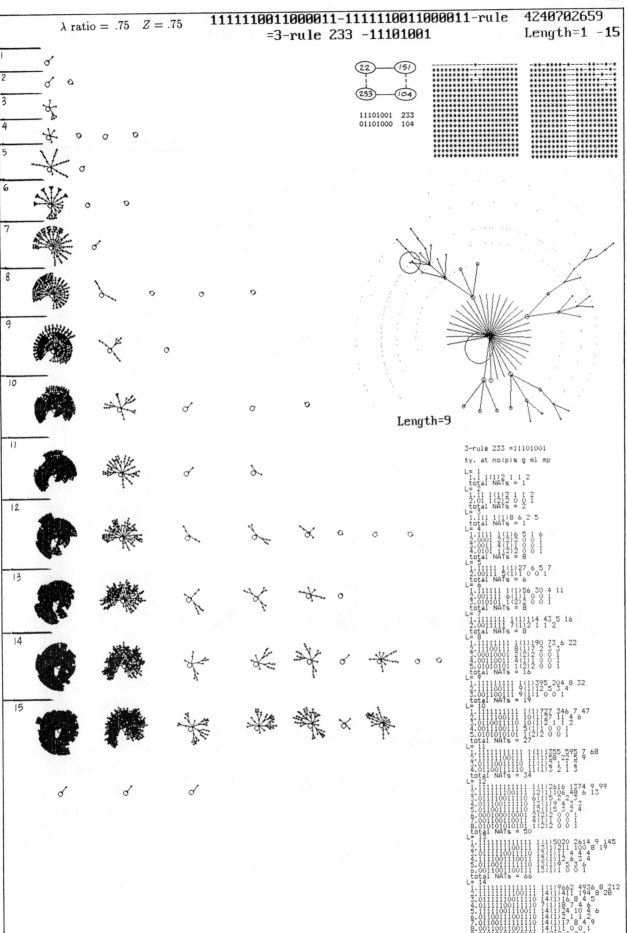

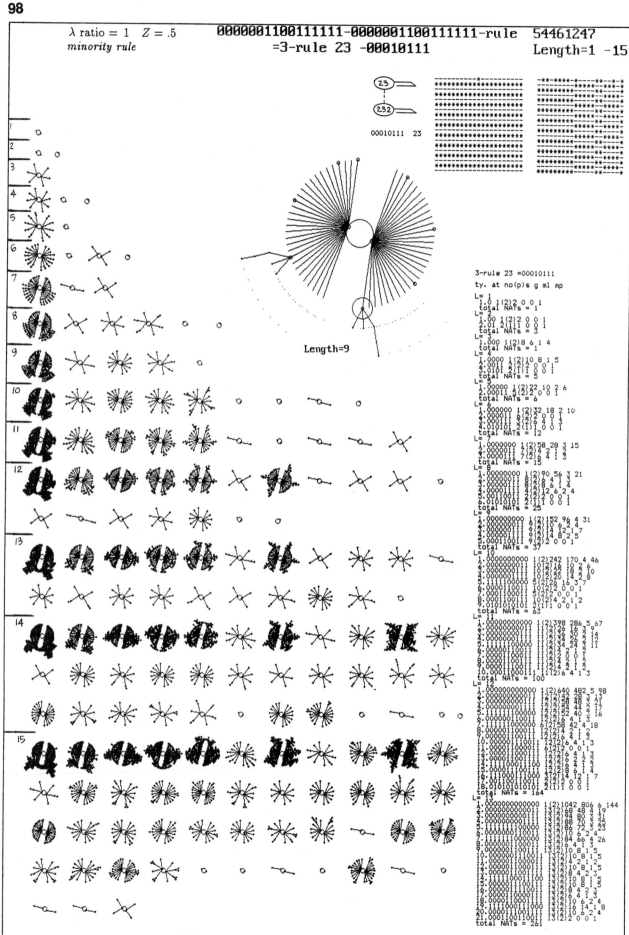

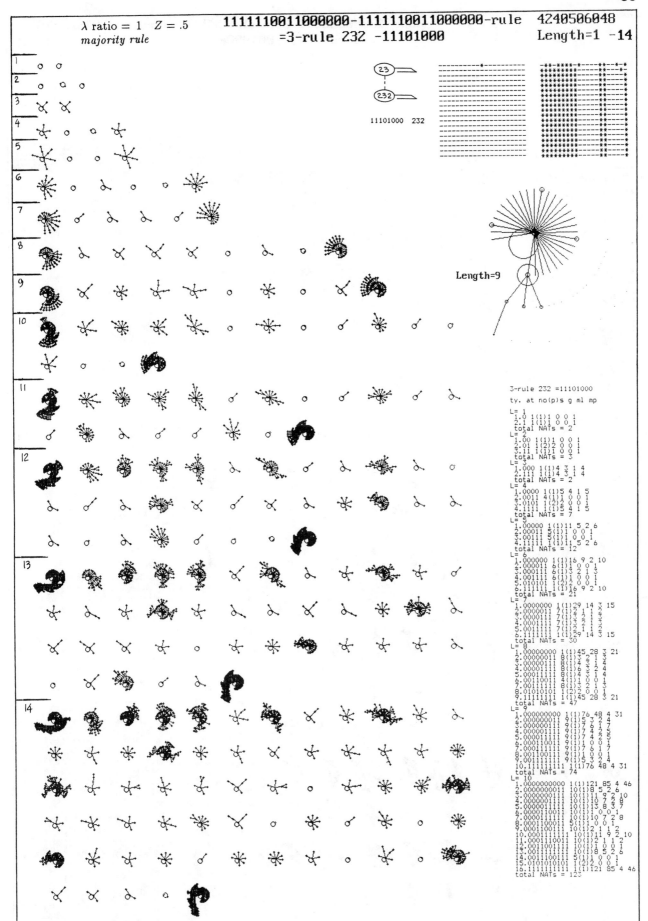

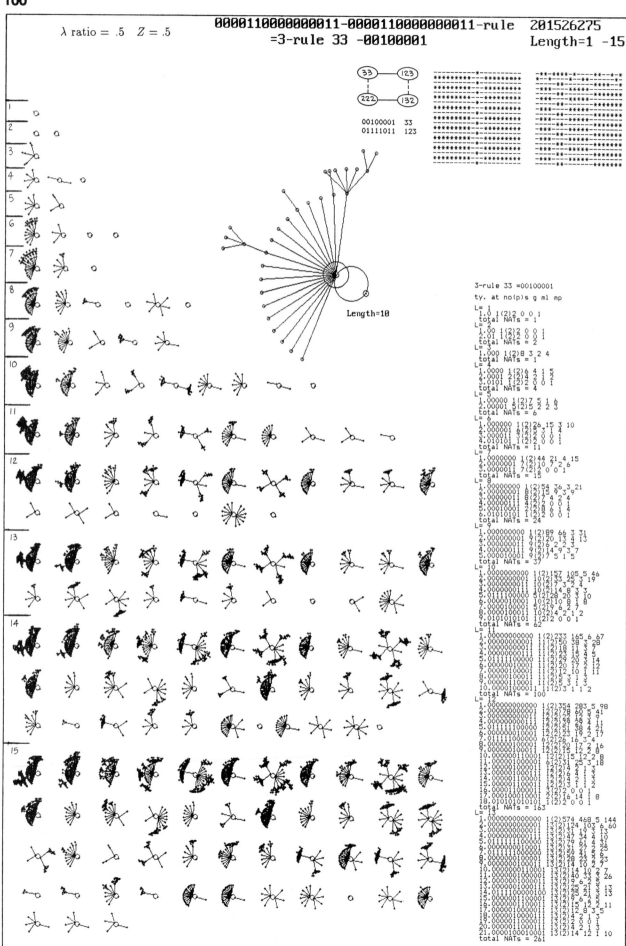

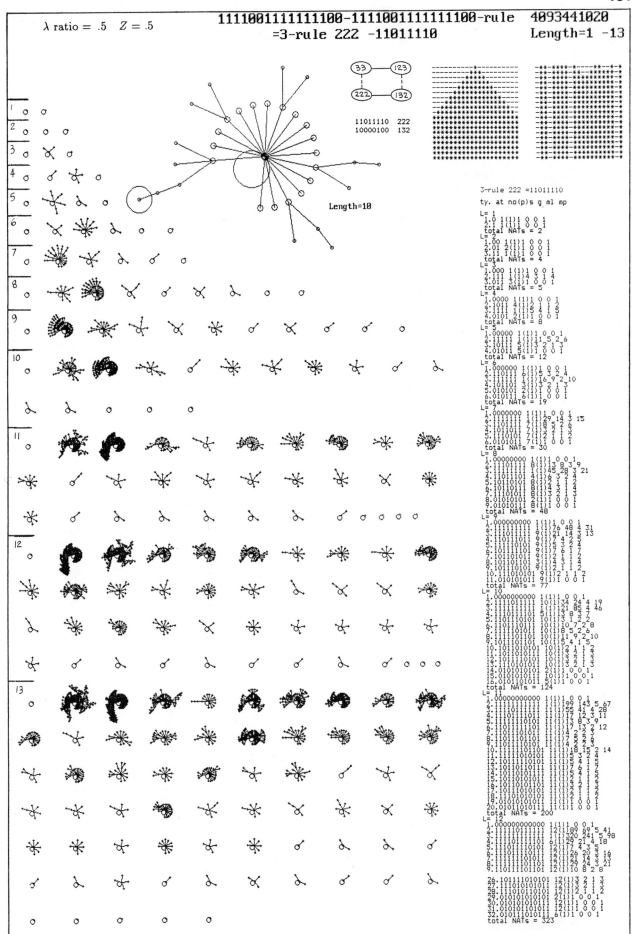

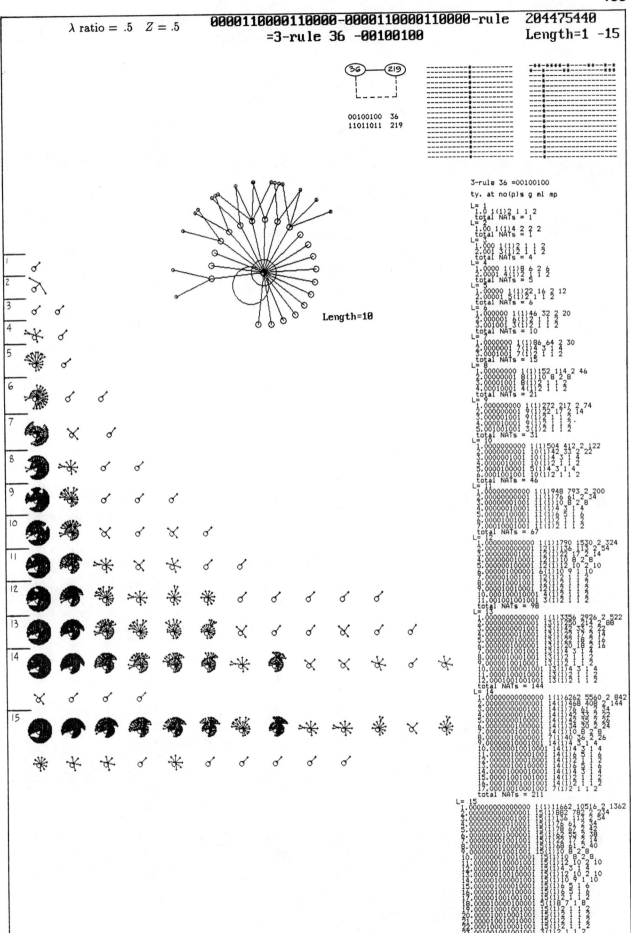

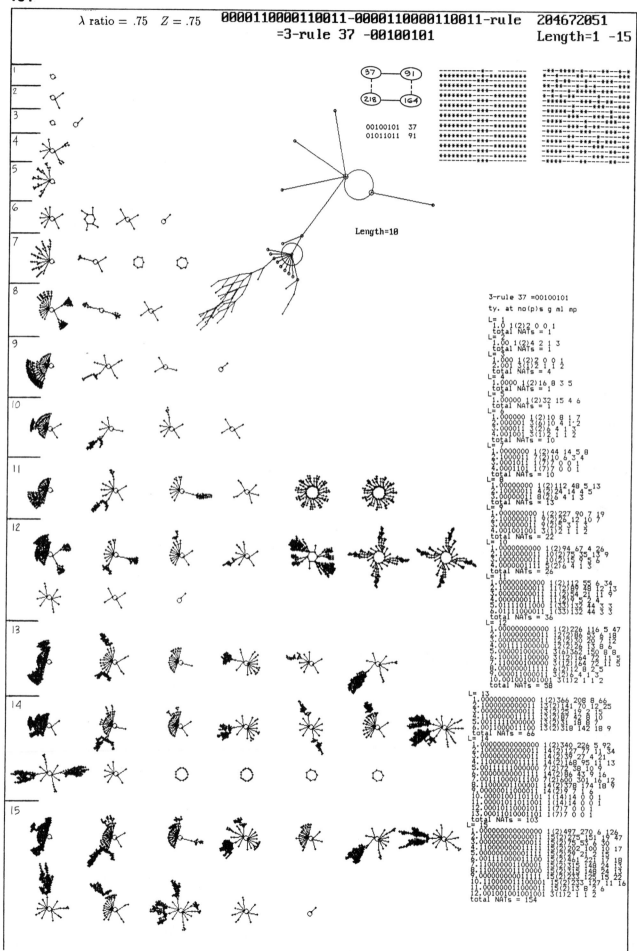

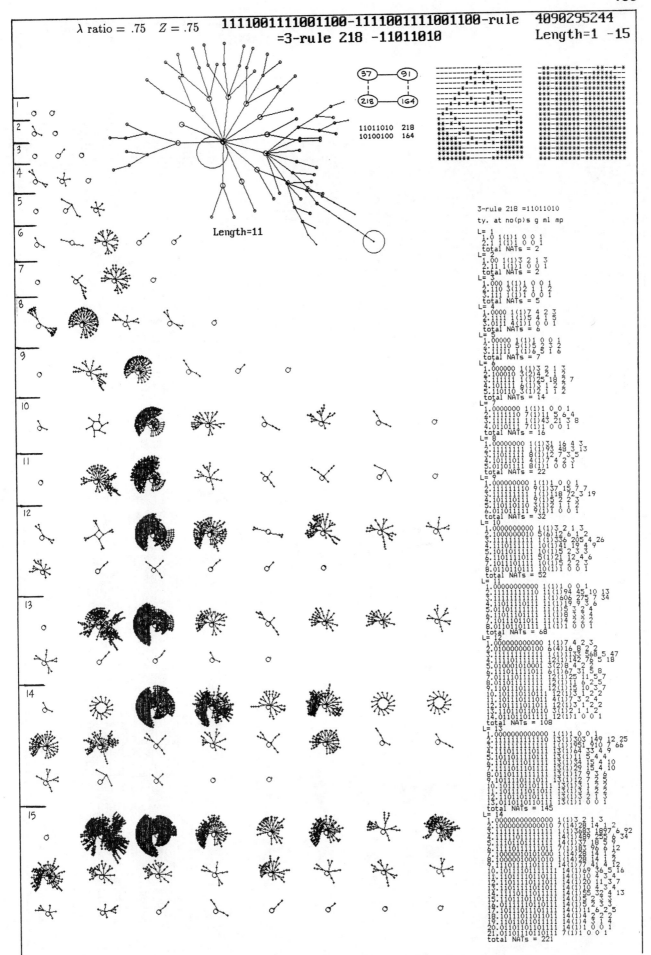

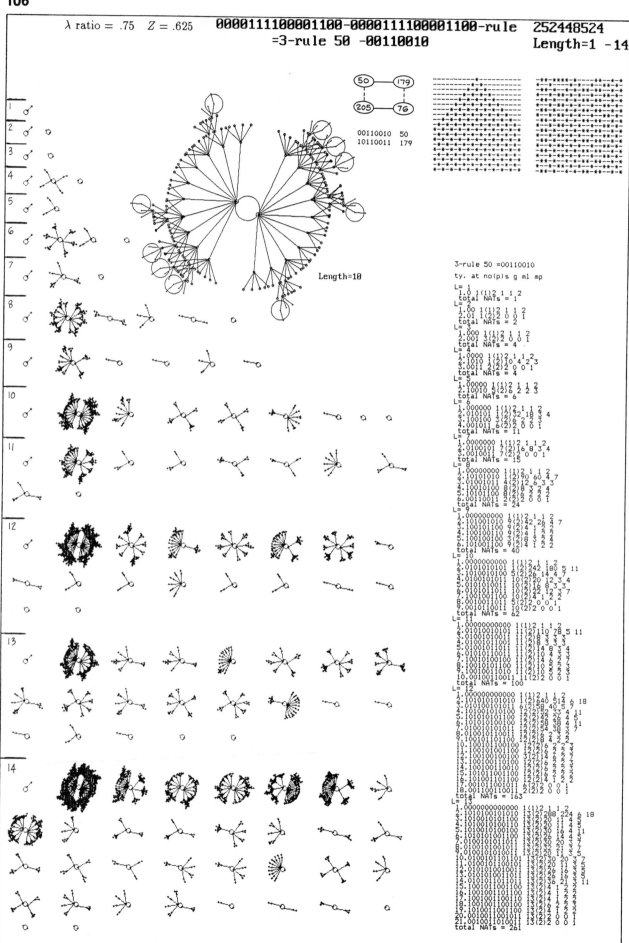

107

λ ratio = .75   Z = .625   1111000011110011-1111000011110011-rule   4042518771
                             =3-rule 205 -11001101   Length=1 -14

11001101 205
01001100   76

all lengths

3-rule 205 =11001101

λ ratio = 1   Z = 1   0000111100001111-0000111100001111-rule   252645135
                     =3-rule 51 -00110011   Length=1 -14

all lengths

3-rule 51 =00110011
ty. at no(p)s g ml mp
L= 1
 1.0 1(2)2 0 0 1
  total NATs = 1
L= 2
 1.00 1(2)2 0 0 1
 2.01 1(2)2 0 0 1
  total NATs = 2
L= 3
 1.000 1(2)2 0 0 1
 2.001 3(2)2 0 0 1
  total NATs = 4
L= 4
 1.0000 1(2)2 0 0 1
 2.0001 4(2)2 0 0 1
 3.0011 2(2)2 0 0 1
 4.0101 1(2)2 0 0 1
  total NATs = 8
L= 5
 1.00000 1(2)2 0 0 1
 2.00001 5(2)2 0 0 1
 3.00011 5(2)2 0 0 1
 4.00101 5(2)2 0 0 1
  total NATs = 16
L= 6
 1.000000 1(2)2 0 0 1
 2.000001 6(2)2 0 0 1
 3.000011 6(2)2 0 0 1
 4.000101 6(2)2 0 0 1
 5.000111 3(2)2 0 0 1
 6.001011 6(2)2 0 0 1
 7.001101 6(2)2 0 0 1
 8.010101 1(2)2 0 0 1
  total NATs = 32
L= 7
 1.0000000 1(2)2 0 0 1
 2.0000001 7(2)2 0 0 1
 3.0000011 7(2)2 0 0 1
 4.0000101 7(2)2 0 0 1
 5.0000111 7(2)2 0 0 1
 6.0001011 7(2)2 0 0 1
 7.0001101 7(2)2 0 0 1
 8.0001111 7(2)2 0 0 1
 9.0010011 7(2)2 0 0 1
10.0010101 7(2)2 0 0 1
  total NATs = 64
L= 8
 1.00000000 1(2)2 0 0 1
 2.00000001 8(2)2 0 0 1
 3.00000011 8(2)2 0 0 1
 4.00000101 8(2)2 0 0 1
 5.00000111 8(2)2 0 0 1
 6.00001011 8(2)2 0 0 1
 7.00001101 8(2)2 0 0 1
 8.00001111 4(2)2 0 0 1
 9.00010011 8(2)2 0 0 1
10.00010001 4(2)2 0 0 1
11.00010011 8(2)2 0 0 1
12.00010101 8(2)2 0 0 1
13.00010111 8(2)2 0 0 1
14.00011011 8(2)2 0 0 1
15.00011011 8(2)2 0 0 1
16.00100101 8(2)2 0 0 1
17.00101001 4(2)2 0 0 1
18.00101011 8(2)2 0 0 1
19.00110011 2(2)2 0 0 1
20.01010101 1(2)2 0 0 1
  total NATs = 128
L= 9
 1.000000000 1(2)2 0 0 1
 2.000000001 9(2)2 0 0 1
 3.000000011 9(2)2 0 0 1
 4.000000101 9(2)2 0 0 1
 5.000000111 9(2)2 0 0 1
 6.000001001 9(2)2 0 0 1
 7.000001011 9(2)2 0 0 1
 8.000001101 9(2)2 0 0 1
 9.000001111 9(2)2 0 0 1
10.000010001 9(2)2 0 0 1
11.000010011 9(2)2 0 0 1
12.000010101 9(2)2 0 0 1
13.000010111 9(2)2 0 0 1
14.000011001 9(2)2 0 0 1
15.000011011 9(2)2 0 0 1
16.000011101 9(2)2 0 0 1
17.000011111 9(2)2 0 0 1
18.000100101 9(2)2 0 0 1
19.000100111 9(2)2 0 0 1
20.000101001 9(2)2 0 0 1
21.000101011 9(2)2 0 0 1
22.000101101 9(2)2 0 0 1
23.000110011 9(2)2 0 0 1
24.000110101 9(2)2 0 0 1
25.000110111 9(2)2 0 0 1
26.001001011 3(2)2 0 0 1
27.001001101 9(2)2 0 0 1
28.001010101 9(2)2 0 0 1
29.001010111 9(2)2 0 0 1
30.001010101 9(2)2 0 0 1
  total NATs = 256

λ ratio = 1 $Z = 1$

1111000011110000-1111000011110000-rule 4042322160
=3-rule 204 -11001100
Length=1 -13

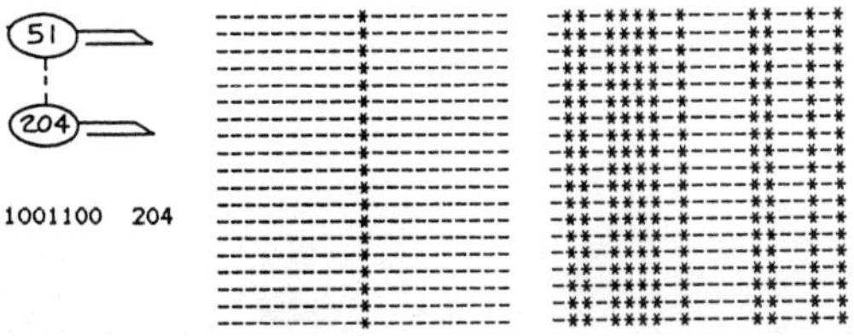

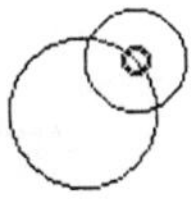

all lengths

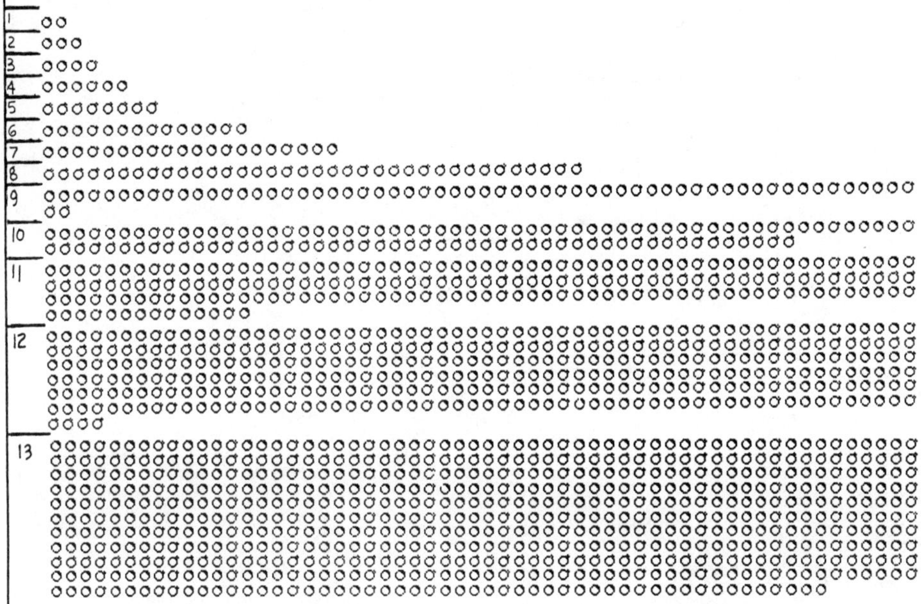


```
3-rule 204 =11001100
ty. at no(p)s g ml mp
L= 1
 1.0 1(1)1 0 0 1
 2.1 1(1)1 0 0 1
 total NATs = 2
L= 2
 1.00 1(1)1 0 0 1
 2.01 2(1)1 0 0 1
 3.11 1(1)1 0 0 1
 total NATs = 4
L= 3
 1.000 1(1)1 0 0 1
 2.001 3(1)1 0 0 1
 3.011 3(1)1 0 0 1
 4.111 1(1)1 0 0 1
 total NATs = 8
L= 4
 1.0000 1(1)1 0 0 1
 2.0001 4(1)1 0 0 1
 3.0011 4(1)1 0 0 1
 4.0101 2(1)1 0 0 1
 5.0111 4(1)1 0 0 1
 6.1111 1(1)1 0 0 1
 total NATs = 16
L= 5
 1.00000 1(1)1 0 0 1
 2.00001 5(1)1 0 0 1
 3.00011 5(1)1 0 0 1
 4.00101 5(1)1 0 0 1
 5.00111 5(1)1 0 0 1
 6.01011 5(1)1 0 0 1
 7.01111 5(1)1 0 0 1
 8.11111 1(1)1 0 0 1
 total NATs = 32
L= 6
 1.000000 1(1)1 0 0 1
 2.000001 6(1)1 0 0 1
 3.000011 6(1)1 0 0 1
 4.000101 6(1)1 0 0 1
 5.000111 6(1)1 0 0 1
 6.001001 3(1)1 0 0 1
 7.001011 6(1)1 0 0 1
 8.001101 6(1)1 0 0 1
 9.001111 6(1)1 0 0 1
 10.010101 2(1)1 0 0 1
 11.010111 6(1)1 0 0 1
 12.011011 3(1)1 0 0 1
 13.011111 6(1)1 0 0 1
 14.111111 1(1)1 0 0 1
 total NATs = 64
L= 7
 1.0000000 1(1)1 0 0 1
 2.0000001 7(1)1 0 0 1
 3.0000011 7(1)1 0 0 1
 4.0000101 7(1)1 0 0 1
 5.0000111 7(1)1 0 0 1
 6.0001001 7(1)1 0 0 1
 7.0001011 7(1)1 0 0 1
 8.0001101 7(1)1 0 0 1
 9.0001111 7(1)1 0 0 1
 10.0010011 7(1)1 0 0 1
 11.0010101 7(1)1 0 0 1
 12.0010111 7(1)1 0 0 1
 13.0011011 7(1)1 0 0 1
 14.0011101 7(1)1 0 0 1
 15.0011111 7(1)1 0 0 1
 16.0101011 7(1)1 0 0 1
 17.0101111 7(1)1 0 0 1
 18.0110111 7(1)1 0 0 1
 19.0111111 7(1)1 0 0 1
 20.1111111 1(1)1 0 0 1
 total NATs = 128
L= 8
 1.00000000 1(1)1 0 0 1
 2.00000001 8(1)1 0 0 1
 3.00000011 8(1)1 0 0 1
 4.00000101 8(1)1 0 0 1
 5.00000111 8(1)1 0 0 1
 6.00001001 8(1)1 0 0 1
 7.00001011 8(1)1 0 0 1
 8.00001101 8(1)1 0 0 1
 9.00001111 8(1)1 0 0 1
 10.00010001 4(1)1 0 0 1
 11.00010011 8(1)1 0 0 1
 12.00010101 8(1)1 0 0 1
 13.00010111 8(1)1 0 0 1
 14.00011001 8(1)1 0 0 1
 15.00011011 8(1)1 0 0 1
 16.00011101 8(1)1 0 0 1
 17.00011111 8(1)1 0 0 1
 18.00100101 8(1)1 0 0 1
 19.00100111 8(1)1 0 0 1
 20.00101011 8(1)1 0 0 1
 21.00101101 8(1)1 0 0 1
 22.00101111 8(1)1 0 0 1
 23.00110011 4(1)1 0 0 1
 24.00110101 8(1)1 0 0 1
 25.00110111 8(1)1 0 0 1
 26.00111011 8(1)1 0 0 1
 27.00111101 8(1)1 0 0 1
 28.00111111 8(1)1 0 0 1
 29.01010101 2(1)1 0 0 1
 30.01010111 8(1)1 0 0 1
 31.01011011 8(1)1 0 0 1
 32.01011111 8(1)1 0 0 1
 33.01101111 8(1)1 0 0 1
 34.01110111 4(1)1 0 0 1
 35.01111111 8(1)1 0 0 1
 36.11111111 1(1)1 0 0 1
 total NATs = 256
```

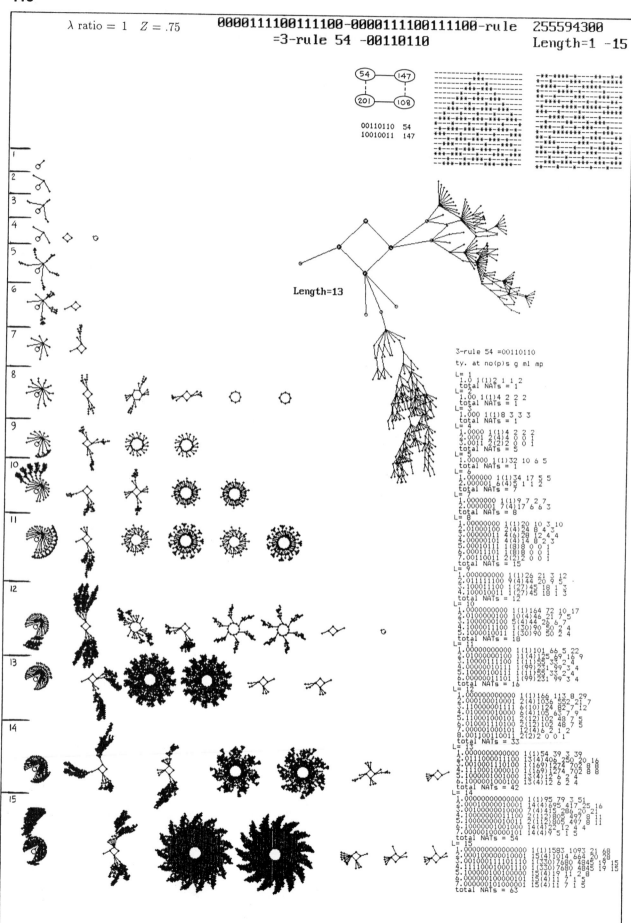

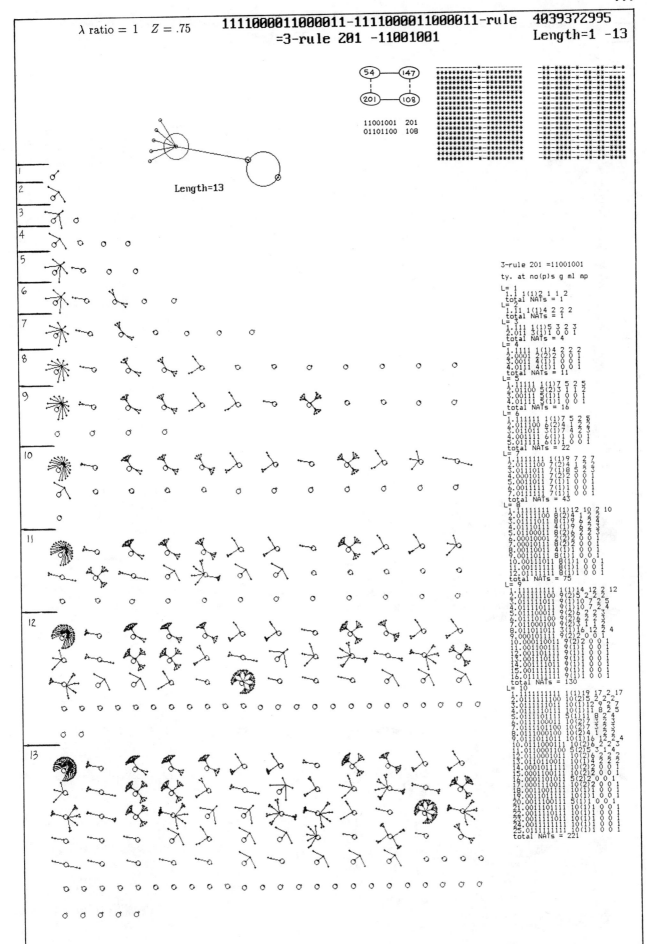

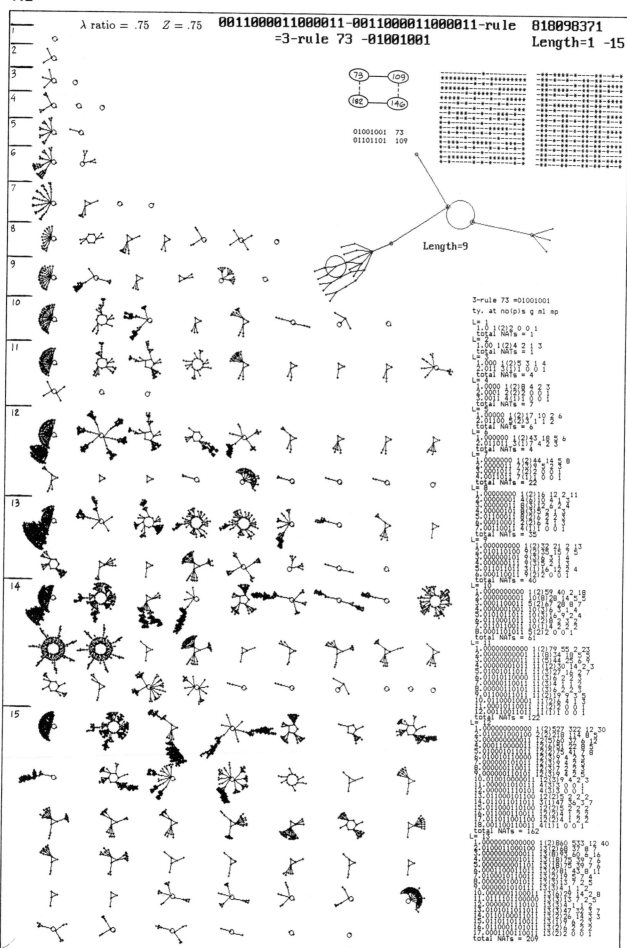

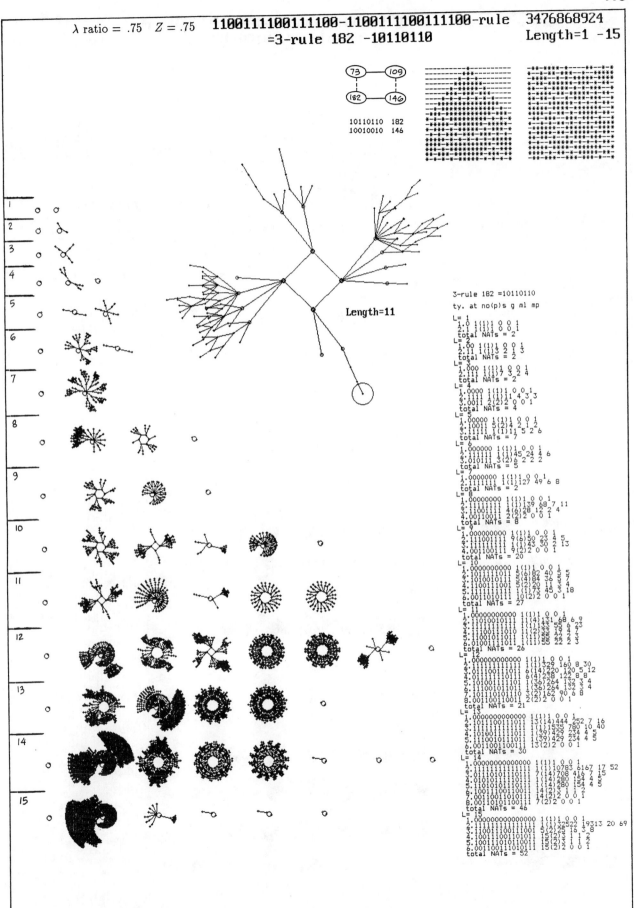

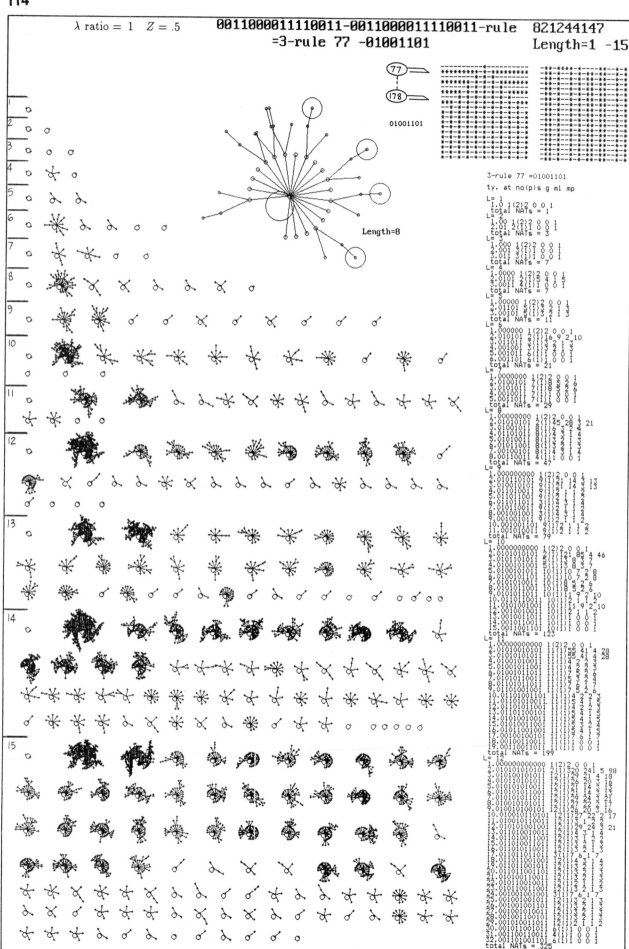

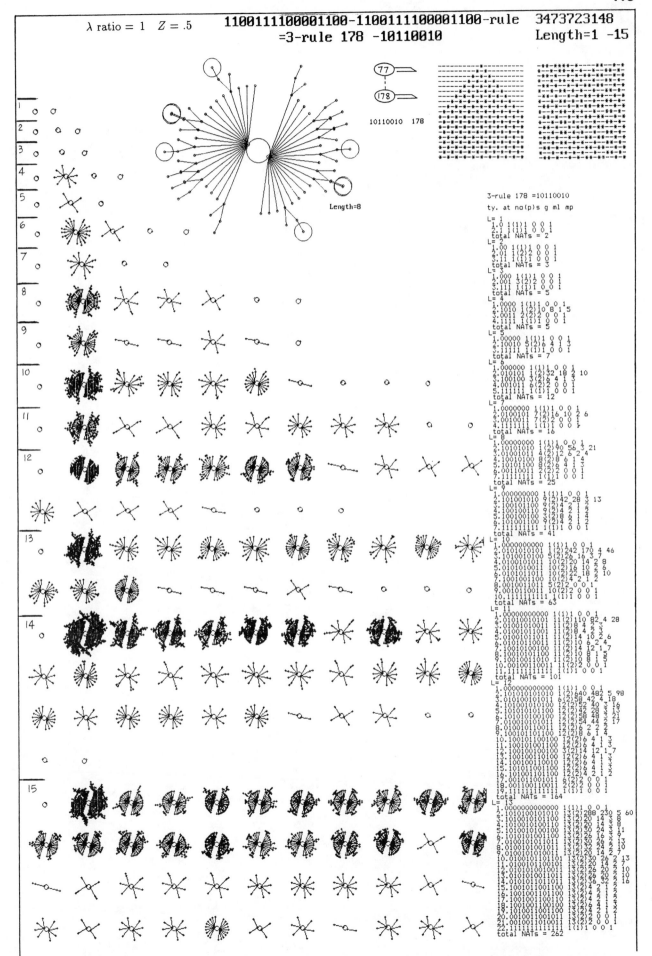

117

λ ratio = 1 $Z = 1$ 0011001111001100-0011001111001100-rule   869020620
                    =3-rule 90 -01011010   Length=1 -14

(90)—(165)

01011010 90
10100101 165

Length=8

3-rule 90 =01011010
ty. at no(p)s g ml mp

L= 1
1.0 1(1)2 1 1 2
total NATs = 1
L= 2
1.00 1(1)4 3 1 4
total NATs = 1
L= 3
1.000 1(1)2 1 1 2
2.110 3(1)2 1 1 2
total NATs = 4
L= 4
1.0000 1(1)16 12 2 4
total NATs = 1
L= 5
1.00000 1(1)2 1 1 2
2.10010 5(3)6 3 1 2
total NATs = 6
L= 6
1.000000 1(1)4 3 1 4
2.100010 3(2)8 6 1 4
3.100101 3(2)8 6 1 4
4.101101 3(1)4 3 1 4
total NATs = 10
L= 7
1.0000000 1(1)2 1 1 2
2.1000010 7(7)14 7 1 2
3.1010101 1(7)14 7 1 2
4.1011100 1(7)14 7 1 2
total NATs = 10
L= 8
1.00000000 1(1)256 192 4 4
total NATs = 1
L= 9
1.000000000 1(1)2 1 1 2
2.100000010 9(7)14 7 1 2
3.100001101 9(7)14 7 1 2
4.100010101 9(7)14 7 1 2
5.100011100 9(7)14 7 1 2
6.110101010 3(1)2 1 1 2
total NATs = 40
L= 10
1.0000000000 1(1)4 3 1 4
2.1000000010 5(6)24 18 1 4
3.1000000111 5(6)24 18 1 4
4.1000001011 5(6)24 18 1 4
5.1000010011 5(6)24 18 1 4
6.1000011011 5(6)24 18 1 4
7.1000011100 5(6)24 18 1 4
8.1001011011 5(6)24 18 1 4
9.1001100101 5(6)24 18 1 4
10.1000110001 5(3)12 9 1 4
total NATs = 46
L= 11
1.00000000000 1(1)2 1 1 2
2.10000000010 11(31)62 31 1 2
3.10000010011 11(31)62 31 1 2
4.10000011100 11(31)62 31 1 2
total NATs = 34
L= 12
1.000000000000 1(1)16 12 2 4
2.100000000010 6(4)64 48 2 4
3.100000000111 6(4)64 48 2 4
4.110000011101 6(4)64 48 2 4
5.100001001011 6(4)64 48 2 4
6.100001011011 6(4)64 48 2 4
7.100001101011 6(4)64 48 2 4
8.010001010101 3(2)32 24 2 4
9.100101010011 6(4)64 48 2 4
10.110001010110 6(4)64 48 2 4
11.110001111011 6(4)64 48 2 4
12.110001111011 6(4)64 48 2 4
13.110011111011 3(2)32 24 2 4
14.110101010110 3(1)16 12 2 4
total NATs = 70
L= 13
1.0000000000000 1(1)2 1 1 2
2.1000000000010 13(63)126 63 1 2
3.1000000010011 13(63)126 63 1 2
4.1000000111100 13(63)126 63 1 2
5.1000001101011 13(63)126 63 1 2
6.1000000111110 13(63)126 63 1 2
total NATs = 66
L= 14
1.00000000000000 1(1)4 3 1 4
2.10000000000010 7(14)56 42 1 4
3.10000000000111 7(14)56 42 1 4
4.10000000001011 7(14)56 42 1 4
5.10000000010011 7(14)56 42 1 4
6.10000000011100 7(14)56 42 1 4
7.10000001001011 7(14)56 42 1 4
8.10000001001100 7(14)56 42 1 4
9.10000010000101 7(14)56 42 1 4
10.10000010010101 7(14)56 42 1 4
11.10000010011011 7(14)56 42 1 4
12.10000010011101 7(14)56 42 1 4
13.10000011000111 7(14)56 42 1 4
14.10000011001011 7(14)56 42 1 4
15.10000011001101 7(14)56 42 1 4
16.10000011010111 7(14)56 42 1 4
17.10000011011101 7(14)56 42 1 4
18.10000011100111 7(14)56 42 1 4
19.10000011101101 7(14)56 42 1 4
20.10000011110101 7(14)56 42 1 4
21.10000100001111 7(14)56 42 1 4
22.10000100010111 7(14)56 42 1 4
23.10000100011101 7(14)56 42 1 4
24.10000100101101 7(14)56 42 1 4
25.10000100110101 7(14)56 42 1 4
26.10000101010101 7(14)56 42 1 4
27.10000101011100 7(14)56 42 1 4
28.10000101011100 7(14)56 42 1 4
29.10000100100111 7(14)56 42 1 4
30.10000110000111 7(14)56 42 1 4
31.10000110001011 7(14)56 42 1 4
32.10000110001101 7(14)56 42 1 4
33.10000110010111 7(14)56 42 1 4
34.10000110011101 7(14)56 42 1 4
35.10000110100111 7(14)56 42 1 4
36.10000110101101 7(14)56 42 1 4
37.10000110110101 7(14)56 42 1 4
38.10000111001101 7(14)56 42 1 4
39.10000111010101 7(14)56 42 1 4
40.10000111011100 7(14)56 42 1 4
41.10000111100101 7(14)56 42 1 4
42.10000111110001 7(14)56 42 1 4
43.10001000100111 7(14)56 42 1 4
44.10001000110101 7(14)56 42 1 4
45.10001001101111 7(14)56 42 1 4
46.10001001111011 7(14)56 42 1 4
47.10001010001111 7(14)56 42 1 4
48.10001010101110 7(14)56 42 1 4
49.10001010110100 7(14)56 42 1 4
50.10001011001111 7(14)56 42 1 4
51.10001011011110 7(14)56 42 1 4
52.10001101001110 1(7)28 21 1 4
total NATs = 298

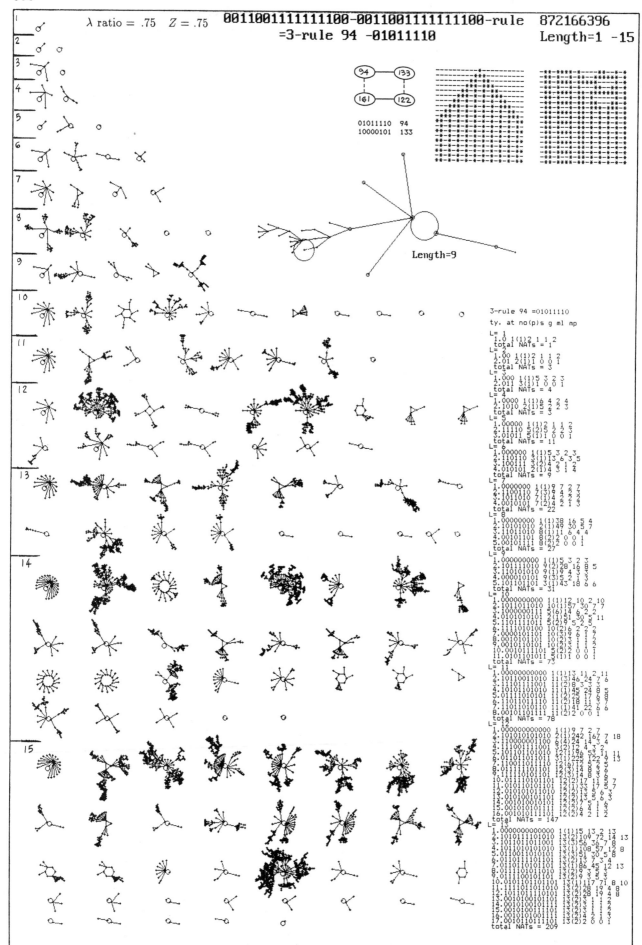

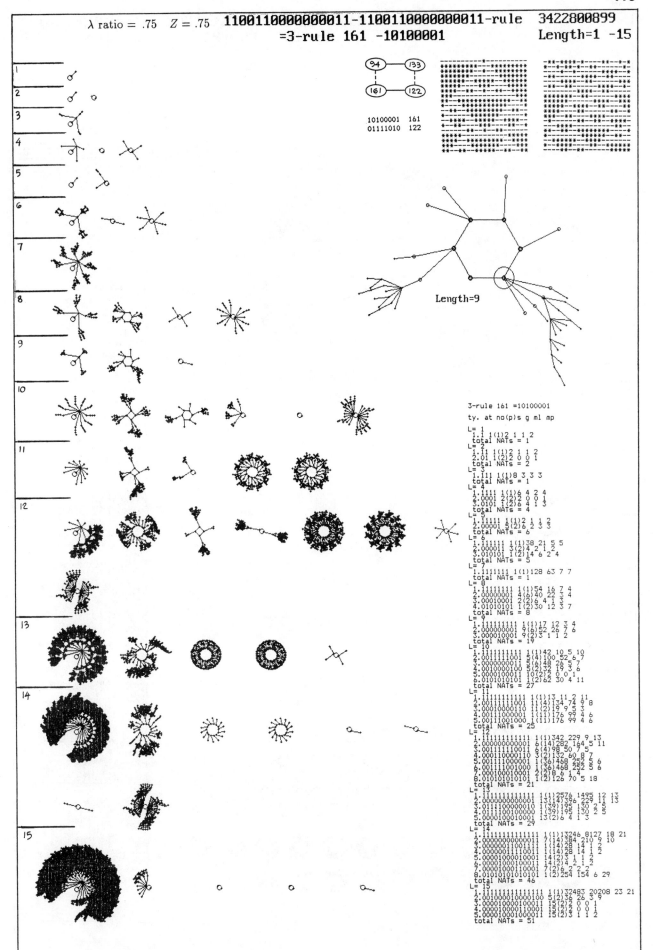

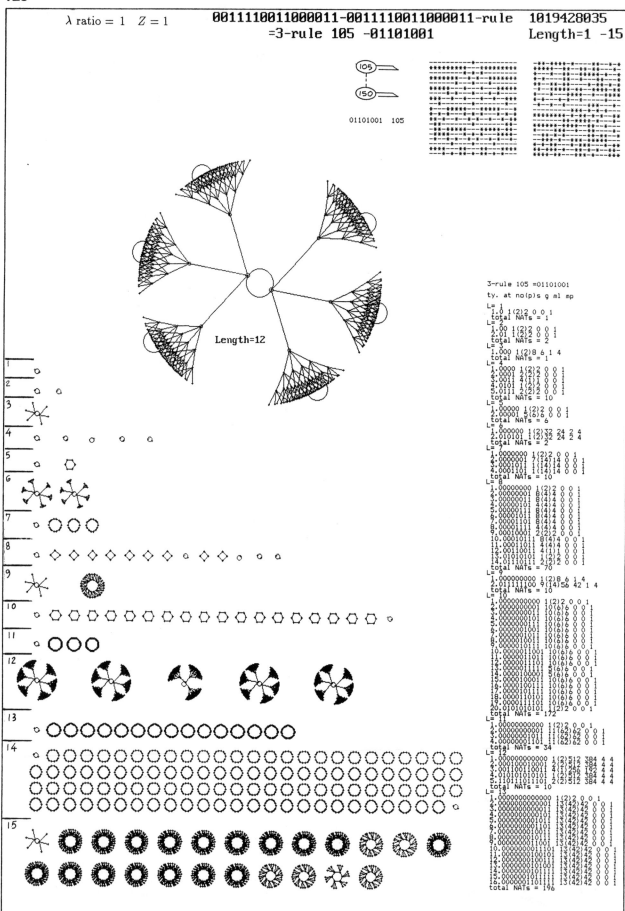

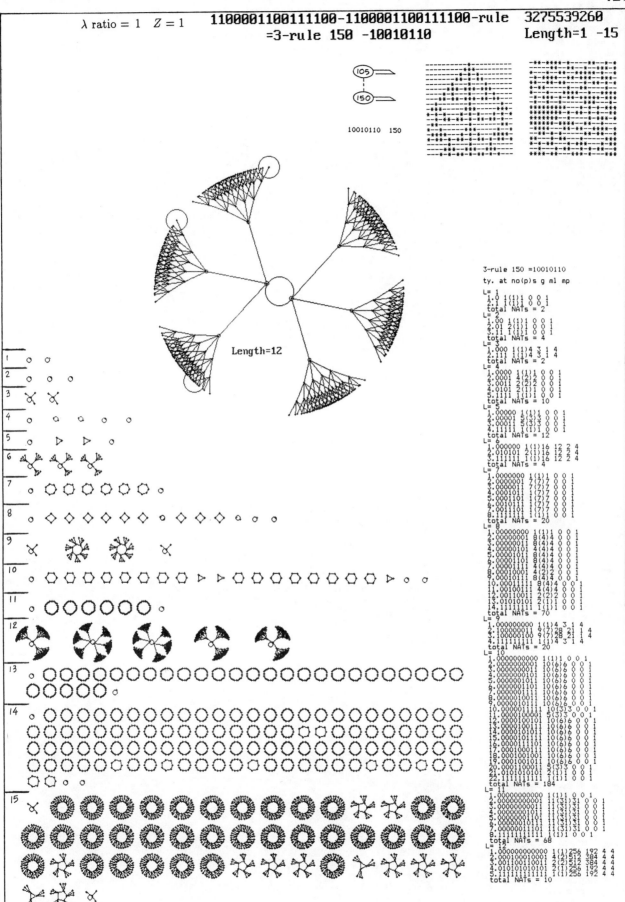

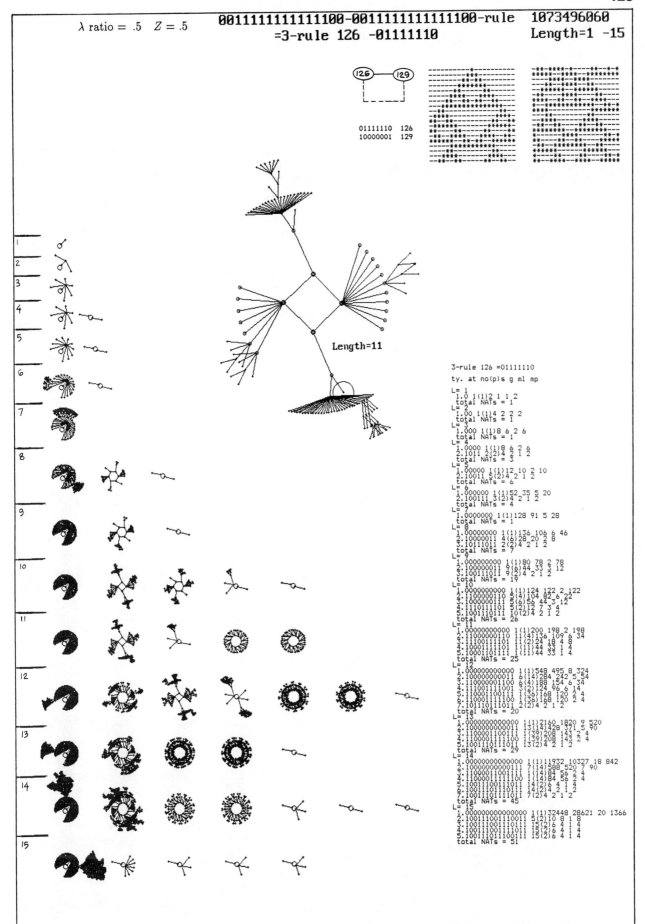

A2.3 n=3 Rules

A2.3.3 Semi-Asymmetric Rule Clusters (see section 3.3.8)

There are no collapsed clusters among the semi-asymmetric rules.

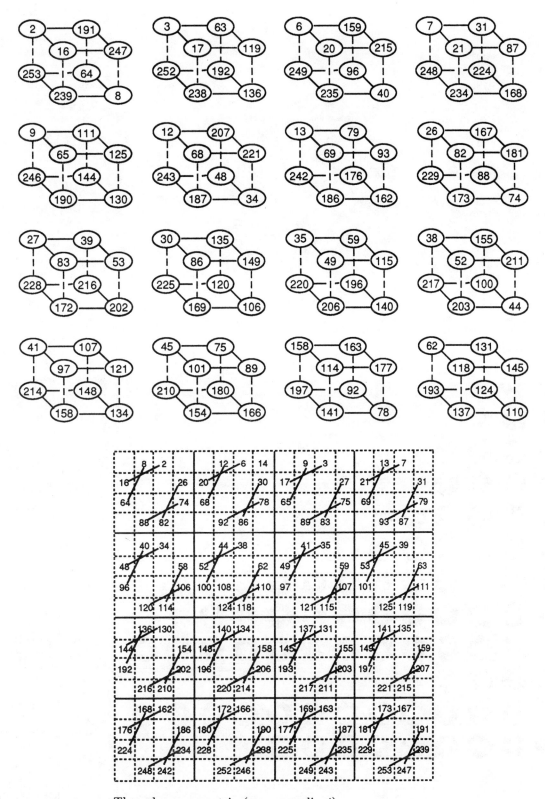

The rule-space matrix (see appendix 4).

126

λ ratio = .25   Z = .25   0000000000001100-0000000000001100-rule 786444
                          =3-rule 2 -00000010   Length=1 -15

```
00000010  2
10111111  191
```

Length=8

```
3-rule 2 =00000010
ty. at no(p)s g ml mp
L= 1
 1.0 1(1)2 1 1 2
  total NATs = 1
L= 2
 1.00 1(1)4 3 1 4
  total NATs = 1
L= 3
 1.000 1(1)5 4 1 5
 2.001 1(3)3 0 0 1
  total NATs = 2
L= 4
 1.0000 1(1)8 7 1 8
 2.0001 1(4)8 4 1 2
  total NATs = 2
L= 5
 1.00000 1(1)12 11 1 12
 2.00001 1(5)20 15 1 4
  total NATs = 2
L= 6
 1.000000 1(1)19 18 1 19
 2.000001 1(6)42 36 1 7
 3.001001 1(3)3 0 0 1
  total NATs = 3
L= 7
 1.0000000 1(1)30 29 1 30
 2.0000001 1(7)84 77 1 12
 3.0001001 1(7)14 7 1 2
  total NATs = 3
L= 8
 1.00000000 1(1)48 47 1 48
 2.00000001 1(8)160 152 1 20
 3.00001001 1(8)32 24 1 4
 4.00010001 1(4)16 12 1 4
  total NATs = 4
L= 9
 1.000000000 1(1)77 76 1 77
 2.000000001 1(9)297 288 1 33
 3.000001001 1(9)63 54 1 7
 4.000010001 1(9)72 63 1 8
 5.001001001 1(3)3 0 0 1
  total NATs = 5
L= 10
 1.0000000000 1(1)124 123 1 124
 2.0000000001 1(10)540 530 1 54
 3.0000001001 1(10)120 110 1 12
 4.0000010001 1(10)140 130 1 14
 5.0000100001 1(5)80 75 1 16
 6.0001001001 1(10)20 10 1 2
  total NATs = 6
L= 11
 1.00000000000 1(1)200 199 1 200
 2.00000000001 1(11)968 957 1 88
 3.00000001001 1(11)220 209 1 20
 4.00000010001 1(11)264 253 1 24
 5.00000100001 1(11)308 297 1 28
 7.00100100001 1(11)44 33 1 4
  total NATs = 7
L= 12
 1.000000000000 1(1)323 322 1 323
 2.000000000001 1(12)1716 1704 1 143
 3.000000001001 1(12)396 384 1 33
 4.000000010001 1(12)480 468 1 40
 5.000000100001 1(12)576 564 1 48
 6.000001000001 1(6)294 288 1 49
 7.000010000001 1(12)84 72 1 7
 8.000100000001 1(12)96 84 1 8
 9.001000000001 1(12)96 84 1 8
10.001001001001 1(4)32 28 1 8
11.001001001001 1(3)3 0 0 1
  total NATs = 11
L= 13
 1.0000000000000 1(1)522 521 1 522
 2.0000000000001 1(13)3016 3003 1 232
 3.0000000001001 1(13)702 689 1 54
 4.0000000010001 1(13)858 845 1 66
 5.0000000100001 1(13)1040 1027 1 80
 6.0000001000001 1(13)1092 1079 1 84
 8.0000100000001 1(13)182 169 1 14
 9.0000100000001 1(13)182 169 1 14
10.0001000000001 1(13)208 195 1 16
11.0010000000001 1(13)208 195 1 16
12.0010010010001 1(13)26 13 1 2
  total NATs = 12
L= 14
 1.00000000000000 1(1)844 843 1 844
 2.00000000000001 1(14)5264 5250 1 376
 4.00000000001001 1(14)1512 1498 1 108
 5.00000000010001 1(14)1848 1834 1 132
 6.00000000100001 1(14)1960 1946 1 140
 7.000000001000001 1(14)280 266 1 20
 8.00000010000001 1(14)1008 1001 1 144
 9.00000100000001 1(14)336 322 1 24
10.00001000000001 1(14)392 378 1 28
11.00010000000001 1(14)392 378 1 28
12.00100000000001 1(14)448 434 1 32
15.00010010010001 1(14)56 42 1 4
16.00010010010001 1(14)56 42 1 4
17.00100100100001 1(7)28 21 1 4
  total NATs = 17
L= 15
 1.000000000000000 1(1)1365 1364 1 1365
 2.000000000000001 1(15)9135 9120 1 609
 3.000000000001001 1(15)2145 2130 1 143
 4.000000000010001 1(15)2640 4625 1 176
 5.000000000100001 1(15)4540 4525 1 176
 6.000000001000001 1(15)3465 3450 1 231
 7.000000010000001 1(15)495 480 1 33
 8.000000100000001 1(15)3600 3585 1 240
 9.000010000000001 1(15)600 585 1 40
10.000100000000001 1(5)90 75 1 6
11.001000000000001 1(15)720 705 1 48
12.010000000000001 1(15)720 705 1 48
13.000010000000001 1(15)720 705 1 48
14.000100010000001 1(15)840 825 1 56
15.001000010000001 1(5)50 35 1 4
16.010000010000001 1(15)840 825 1 56
17.000010010010001 1(5)105 90 1 7
18.001001001001001 1(5)320 315 1 64
19.010010010010001 1(15)120 105 1 8
20.001001001001001 1(15)120 105 1 8
21.010010010010001 1(15)120 105 1 8
22.010010010010001 1(15)120 105 1 8
23.001001001001001 1(3)3 0 0 1
  total NATs = 23
```

λ ratio = .25   Z = .25   **1111111111110011-1111111111110011-rule   4294180851**
                          =3-rule 253 -11111101   Length=1 -15

11111101 253
01000000   64
00001000   8
11101111 239

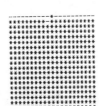 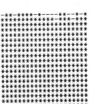

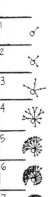

1
2
3
4
5
6
7
8
9
10
11
12
13
14
15

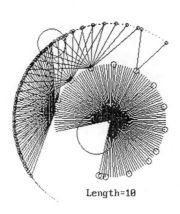

Length=10

3-rule 253 =11111101
ty. at no(p)s g ml mp
L= 1
 1.1 1(1)2 1 1 2
  total NATs = 1
L= 2
 1.11 1(1)4 3 1 4
  total NATs = 1
L= 3
 1.111 1(1)8 4 2 5
  total NATs = 1
L= 4
 1.1111 1(1)16 11 2 8
  total NATs = 1
L= 5
 1.11111 1(1)32 26 2 12
  total NATs = 1
L= 6
 1.111111 1(1)64 54 2 19
  total NATs = 1
L= 7
 1.1111111 1(1)128 113 2 30
  total NATs = 1
L= 8
 1.11111111 1(1)256 235 2 48
  total NATs = 1
L= 9
 1.111111111 1(1)512 481 2 77
  total NATs = 1
L= 10
 1.1111111111 1(1)1024 978 2 124
  total NATs = 1
L= 11
 1.11111111111 1(1)2048 1981 2 200
  total NATs = 1
L= 12
 1.111111111111 1(1)4096 3998 2 323
  total NATs = 1
L= 13
 1.1111111111111 1(1)8192 8048 2 522
  total NATs = 1
L= 14
 1.11111111111111 1(1)16384 16173 2 844
  total NATs = 1
L= 15
 1.111111111111111 1(1)32768 32459 2 1365
  total NATs = 1

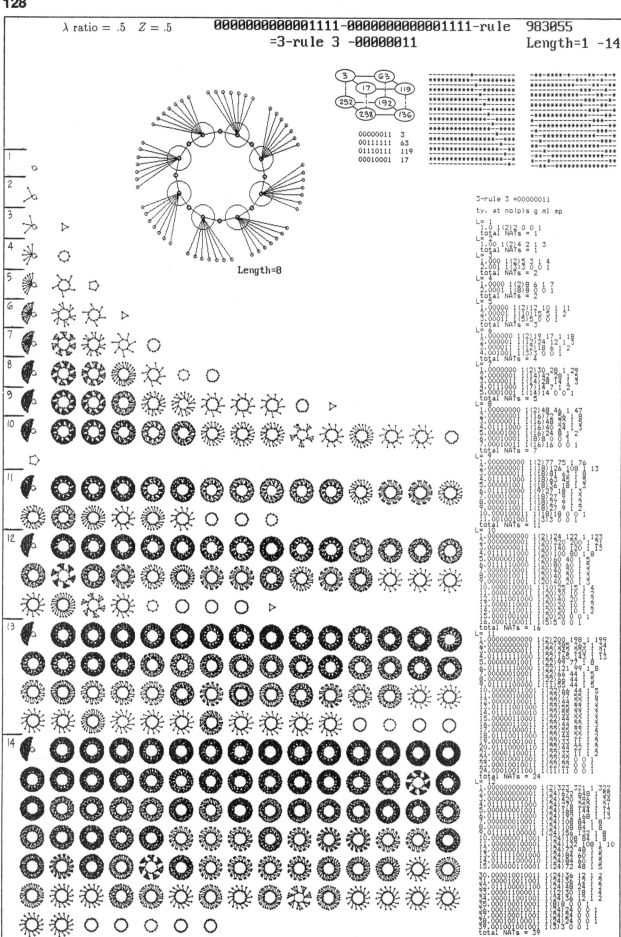

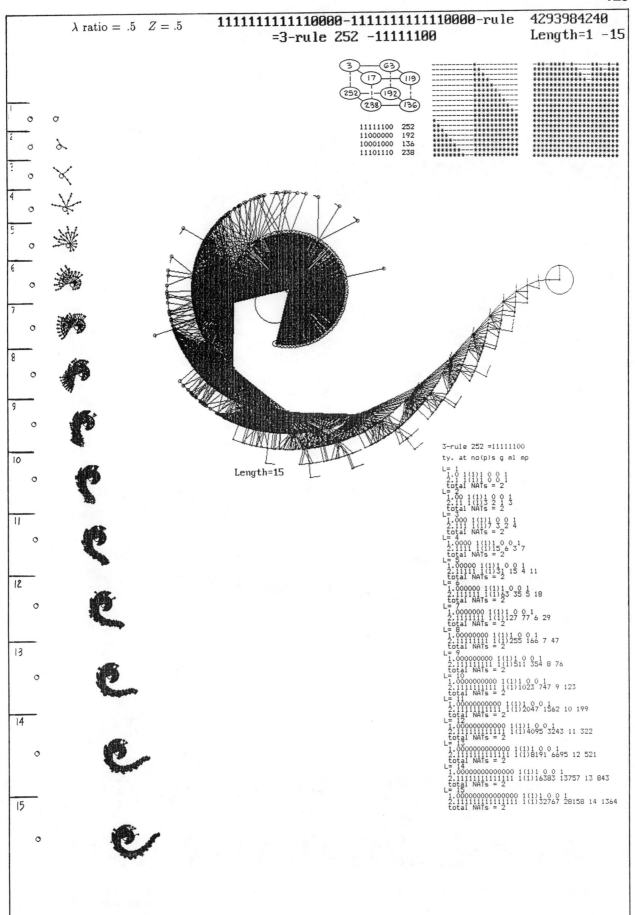

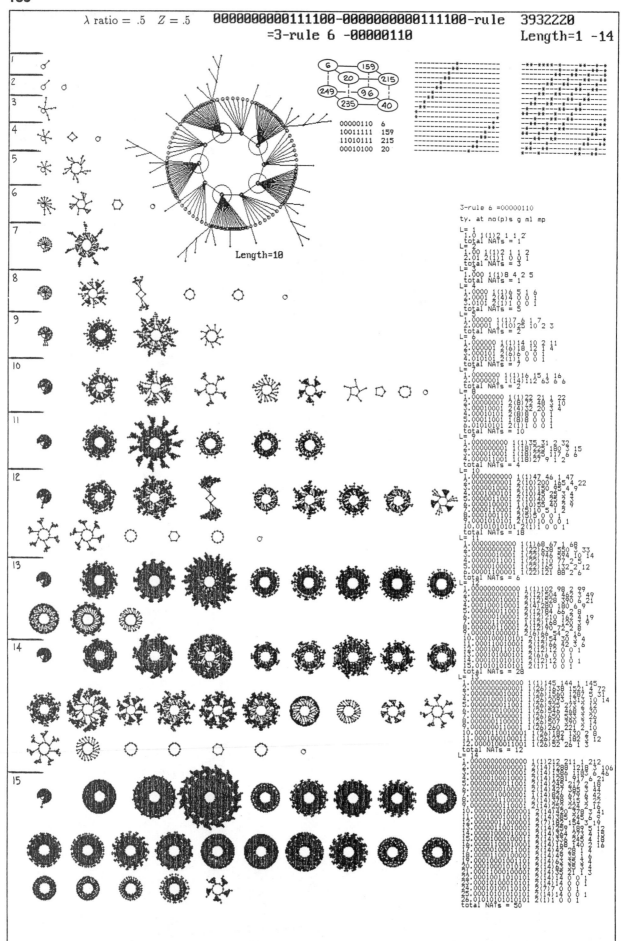

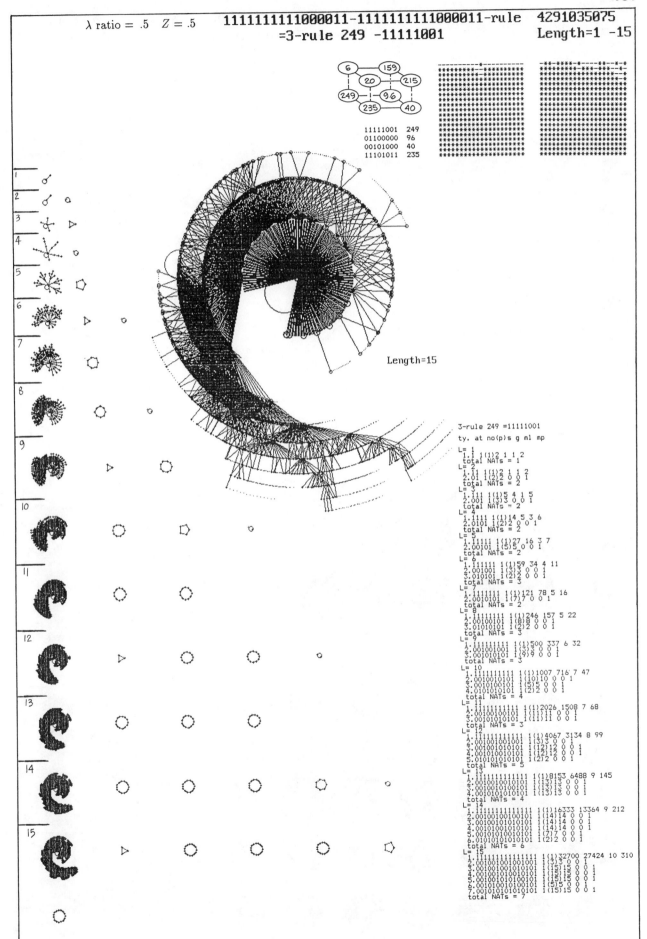

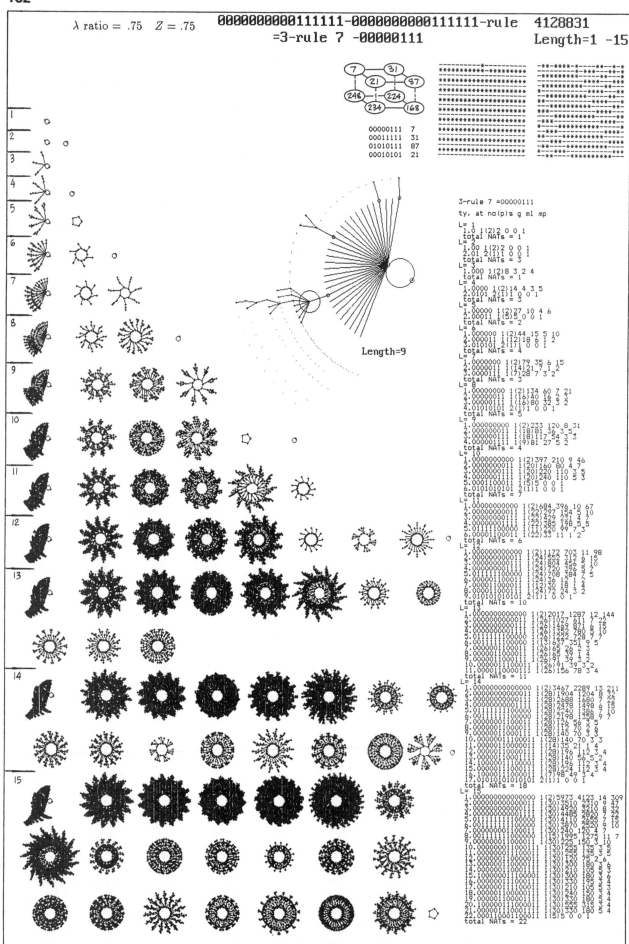

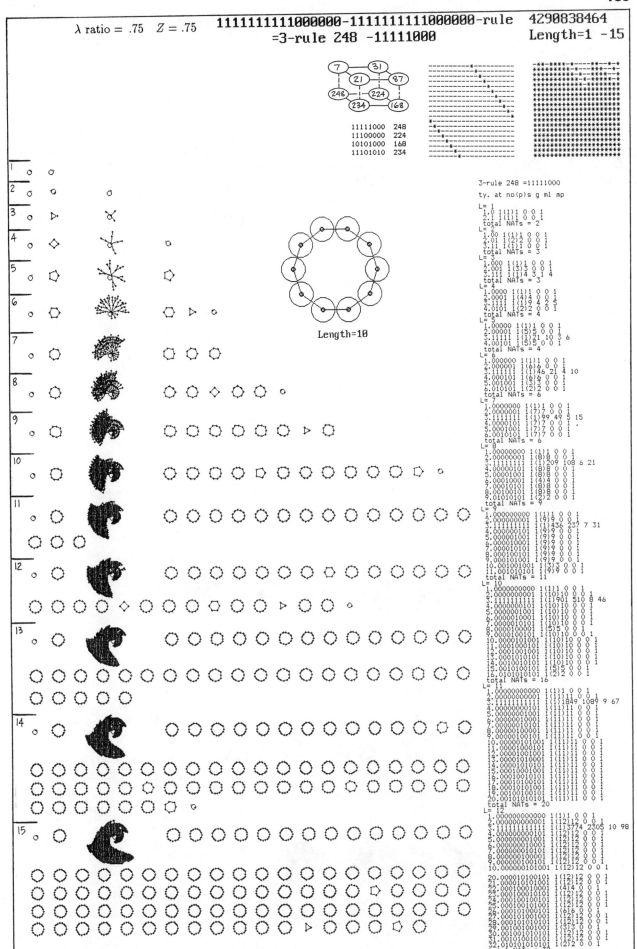

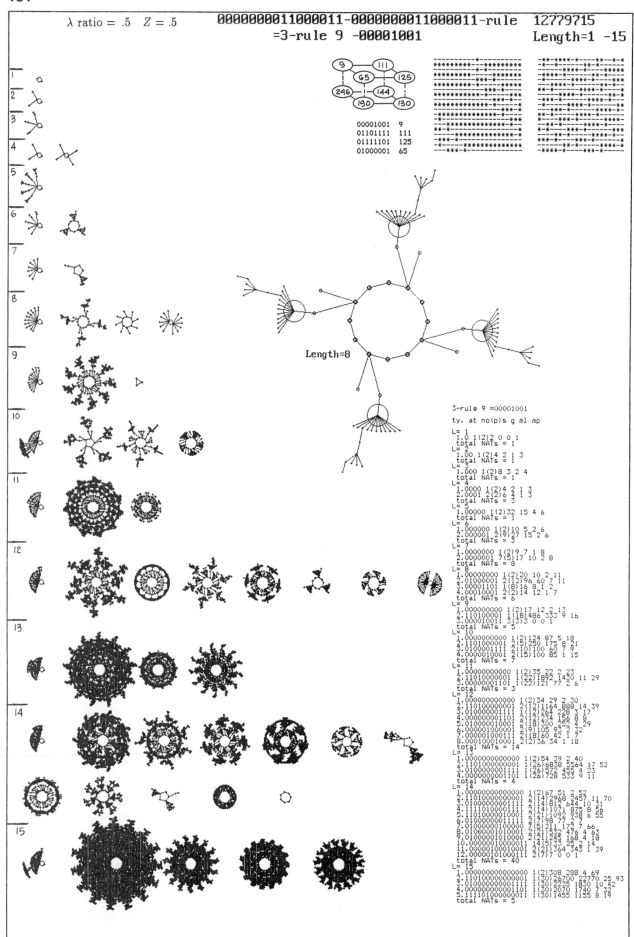

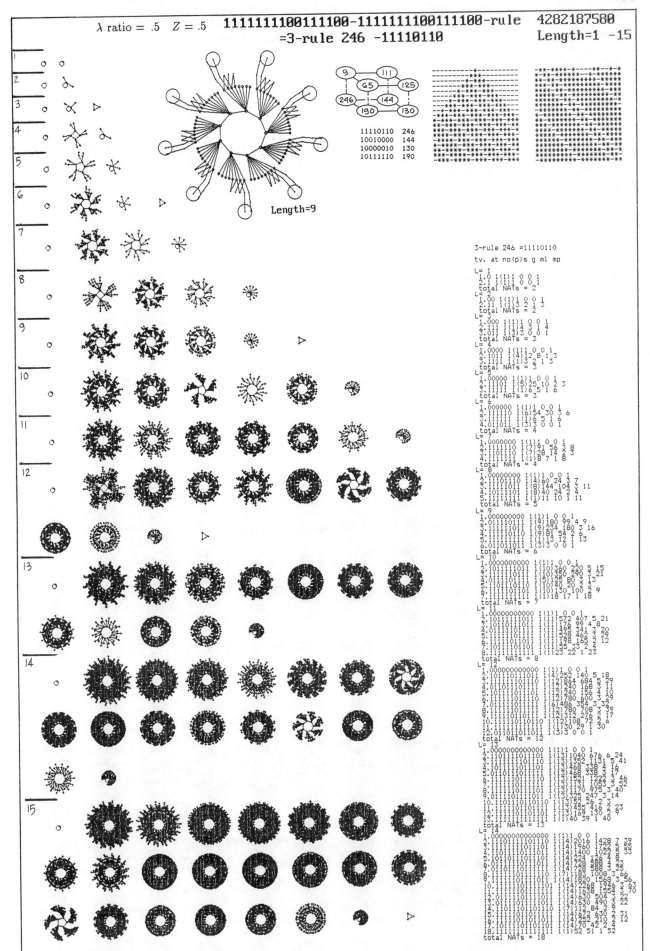

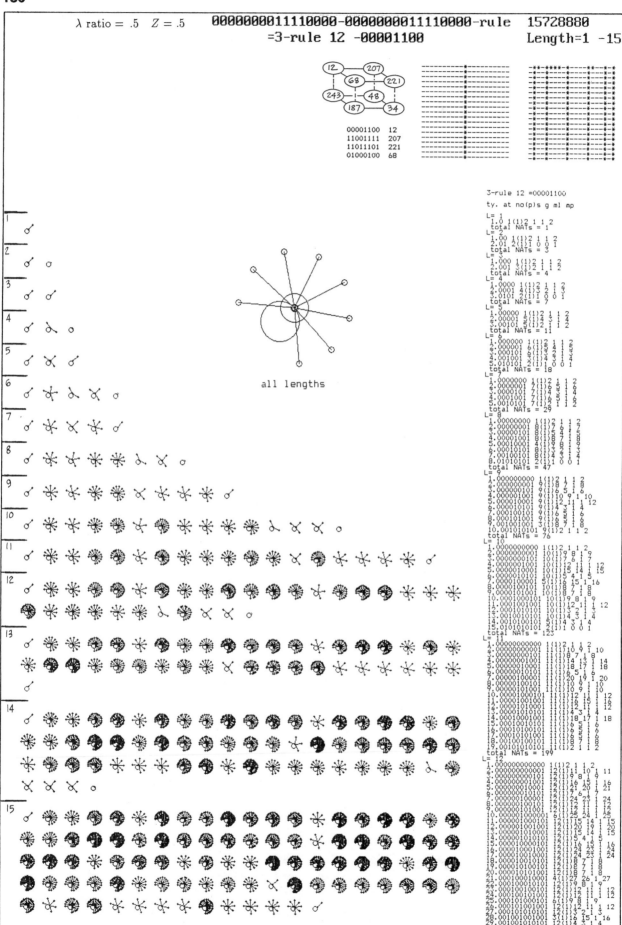

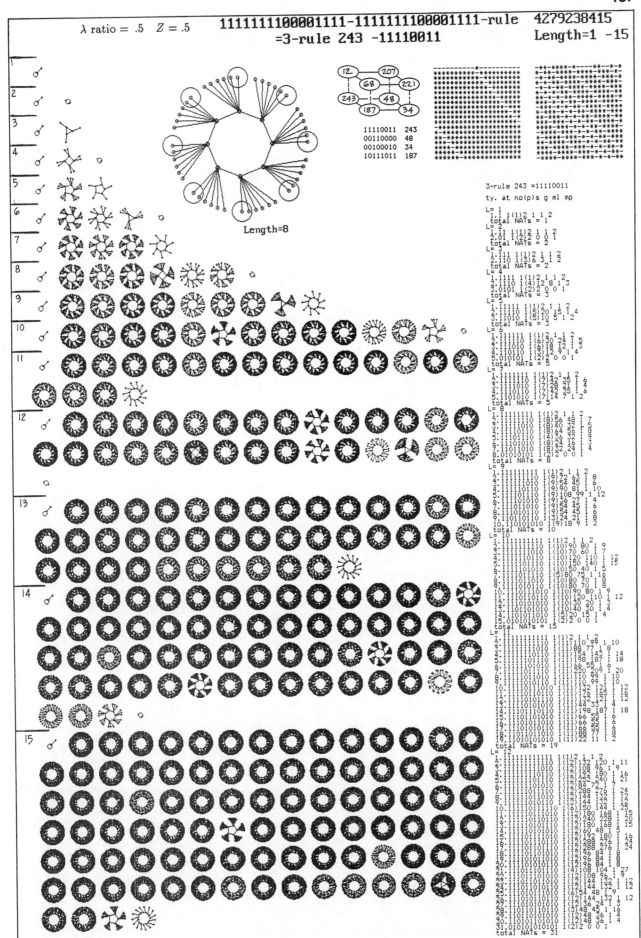

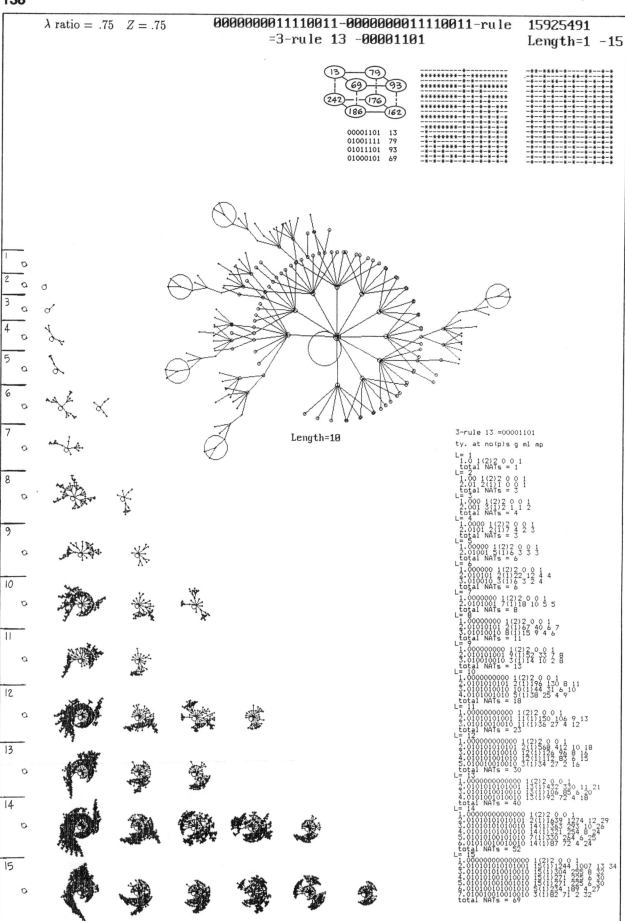

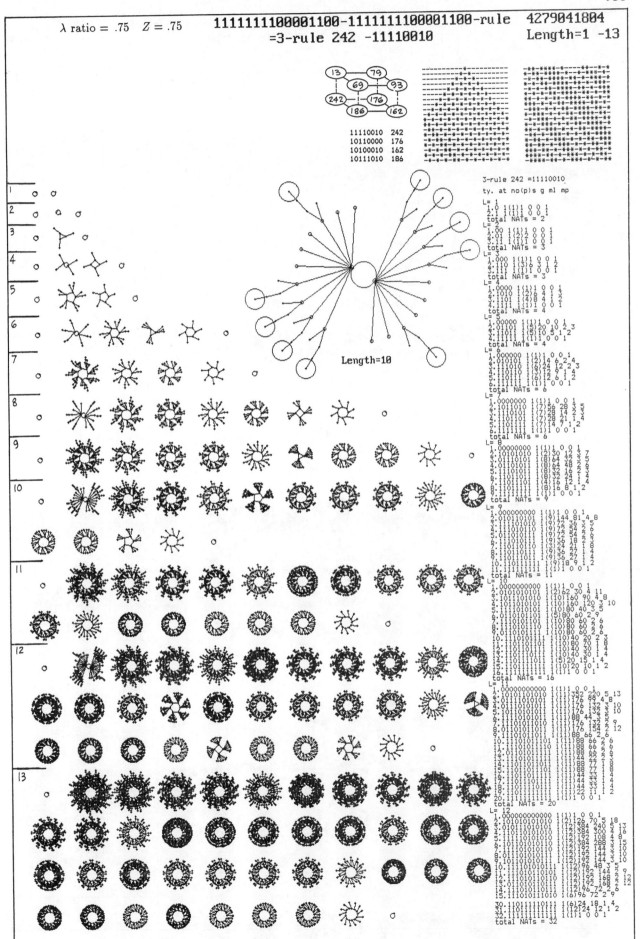

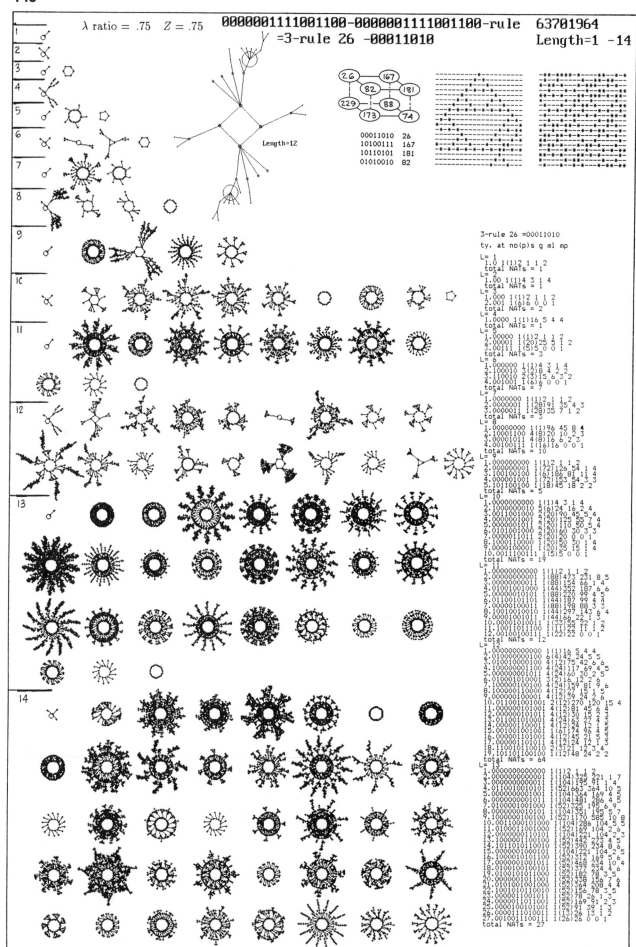

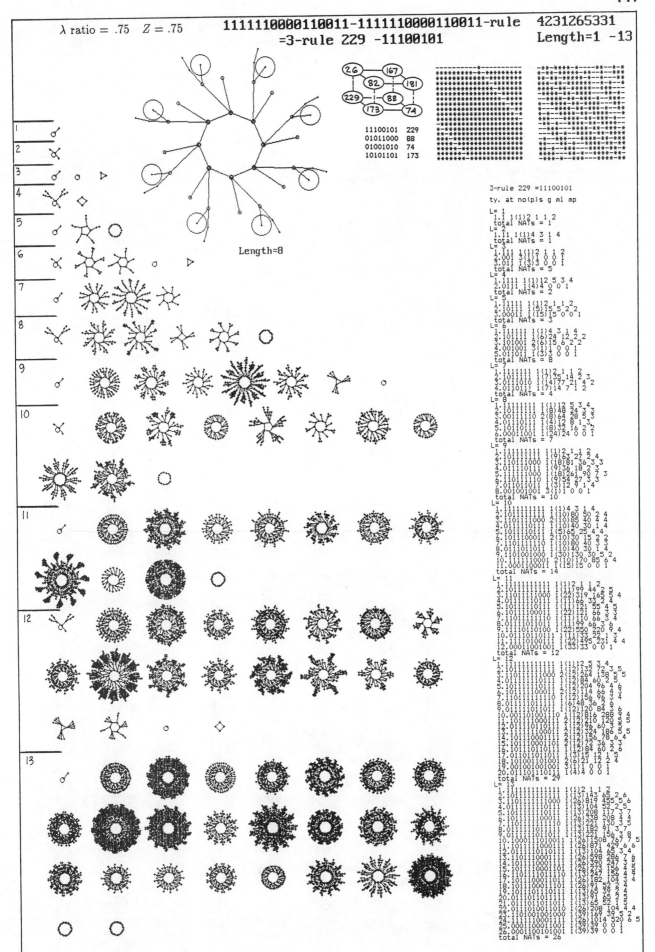

142

λ ratio = 1   Z = .75   00000001111001111-00000001111001111-rule   63898575
=3-rule 27 -00011011   Length=1 -13

```
00011011  27
00100111  39
00110101  53
01010011  83
```

Length=7

3-rule 27 =00011011
ty. at no(p)s g ml mp
L= 1
 1.0   1(2)2 0 0 1
  total NATs = 1
L= 2
 1.00   1(2)4 2 1 3
  total NATs = 1
L= 3
 1.000   1(2)2 0 0 1
 2.001   1(6)6 0 0 1
  total NATs = 2
L= 4
 1.0000 1(2)4 2 1 3
 2.0001 1(8)12 4 1 2
  total NATs = 2
L= 5
 1.00000 1(2)2 0 0 1
 2.00001 1(10)25 10 2 2
 3.00111 1(5)5 0 0 1
  total NATs = 3
L= 6
 1.000000 1(2)4 2 1 3
 2.000001 1(12)24 12 1 2
 3.111000 1(12)24 12 1 3
 4.001001 1(6)6 0 0 1
 5.001011 2(3)3 0 0 1
  total NATs = 6
L= 7
 1.0000000 1(2)2 0 0 1
 2.0000001 1(14)42 21 2 3
 3.1111100 1(7)21 14 1 3
 4.0000011 1(14)21 7 1 2
 5.0001001 1(14)42 14 3 2
  total NATs = 5
L= 8
 1.00000000 1(2)4 2 1 3
 2.00000001 1(16)40 24 1 3
 3.11111100 1(16)40 24 1 4
 4.00000111 1(16)40 24 1 3
 5.00010001 1(16)80 32 4 2
 6.00100001 1(8)20 12 1 4
 7.11100100 1(16)32 8 2 2
  total NATs = 7
L= 9
 1.000000000 1(2)2 0 0 1
 2.000000001 1(18)63 36 2 4
 3.111111000 1(18)63 45 1 4
 4.000000111 1(18)54 18 1 3
 5.000001001 1(18)72 36 3 2
 6.000010001 1(18)72 45 2 3
 7.111100001 1(18)63 45 2 3
 8.111110001 1(9)18 9 1 1
 9.111110100 1(18)72 36 4 2
10.000100111 1(18)27 9 1 2
11.001001001 1(6)6 0 0 1
12.001100101 1(18)18 0 0 1
  total NATs = 12
L= 10
 1.0000000000 1(2)4 2 1 3
 2.0000000001 1(20)60 40 1 4
 3.1111111000 1(20)60 40 1 3
 4.0000000111 1(20)60 40 1 4
 5.0000001001 1(20)60 40 1 4
 6.0000010001 1(20)60 40 1 4
 7.1111100001 1(20)60 40 1 4
 8.0000100001 1(10)65 30 5 2
 9.0001100001 1(20)80 30 3 4
10.0111100001 1(20)50 30 1 4
11.0001001001 1(20)50 30 1 4
12.0001001001 1(20)50 30 1 4
13.1111100100 1(20)50 20 0 1
14.0011000111 1(20)120 40 9 2
15.0011100001 1(20)120 40 9 2
16.0011100111 1(5)5 0 0 1
  total NATs = 16
L= 11
 1.00000000000 1(2)2 0 0 1
 2.00000000001 1(22)88 55 2 5
 3.11111110000 1(22)88 55 1 5
 4.00000000111 1(22)55 15 1 3
 5.00000001001 1(22)110 66 3 3
 6.00000010001 1(11)44 22 2 4
 7.11111100001 1(22)110 77 2 4
 8.00001100001 1(22)110 77 2 4
10.10011110100 1(22)110 77 2 4
11.01111111000 1(22)44 22 2 4
12.00001100011 1(22)110 77 2 4
13.11111100110 1(22)110 66 5 3
14.00001000110 1(22)110 66 3 3
15.11110011000 1(22)99 79 1 6
16.00010001011 1(22)88 22 6 2
17.00000100011 1(22)99 88 2 6
18.11111001110 1(22)99 88 2 6
19.00001110111 1(22)44 22 2 4
20.00001110111 1(22)55 33 1 4
21.01110001001 1(22)110 66 3 4
22.00011001001 1(22)110 66 3 4
23.00011001010 1(22)44 22 2 2
24.11110010001 1(11)44 22 2 4
  total NATs = 24
L= 12
 1.000000000000 1(2)4 2 1 3
 2.000000000001 1(24)84 60 1 4
 3.111111110000 1(24)84 60 1 5
 4.000000000111 1(24)84 60 1 4
 5.000000001001 1(24)168 96 6 4
 6.000000010001 1(24)84 60 1 4
 7.111111100001 1(24)84 60 1 4
 8.000001100001 1(24)84 60 2 4
 9.111111000001 1(24)84 60 2 4
10.011111100100 1(24)192 120 6 4
11.011111100100 1(24)168 120 2 3
12.000011100001 1(24)168 120 3 6
13.000001011100 1(24)84 60 2 3
14.000001011100 1(24)96 24 3 4
15.000000111011 1(24)96 24 3 4
16.111111111000 1(24)96 24 3 4
17.111001011000 1(24)96 24 3 4
18.111110001100 1(24)192 96 5 2
19.111000111100 1(24)84 60 1 4
20.000011010111 1(24)84 60 1 4
35.001110010010 1(24)192 96 5 4
36.000010110111 1(24)72 24 3 2
37.001100111001 1(24)72 24 3 2
38.111001110001 1(24)60 48 1 2
39.000100011001 1(6)6 0 0 1
40.000100011011 1(24)12 6 0 1
41.000101011011 1(2)12 0 0 1
42.001011001011 2(3)3 0 0 1
  total NATs = 44
```
```

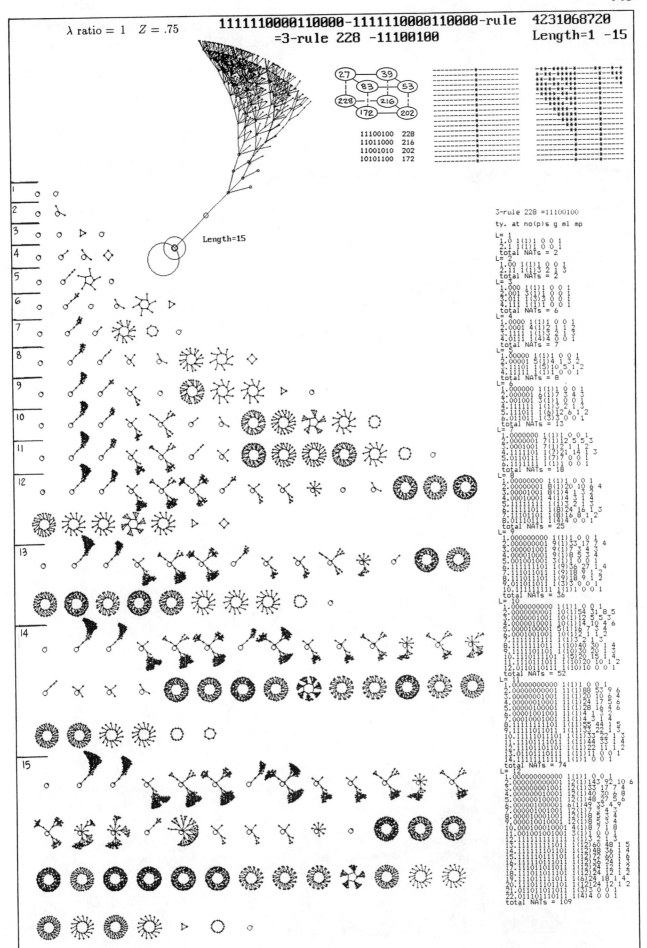

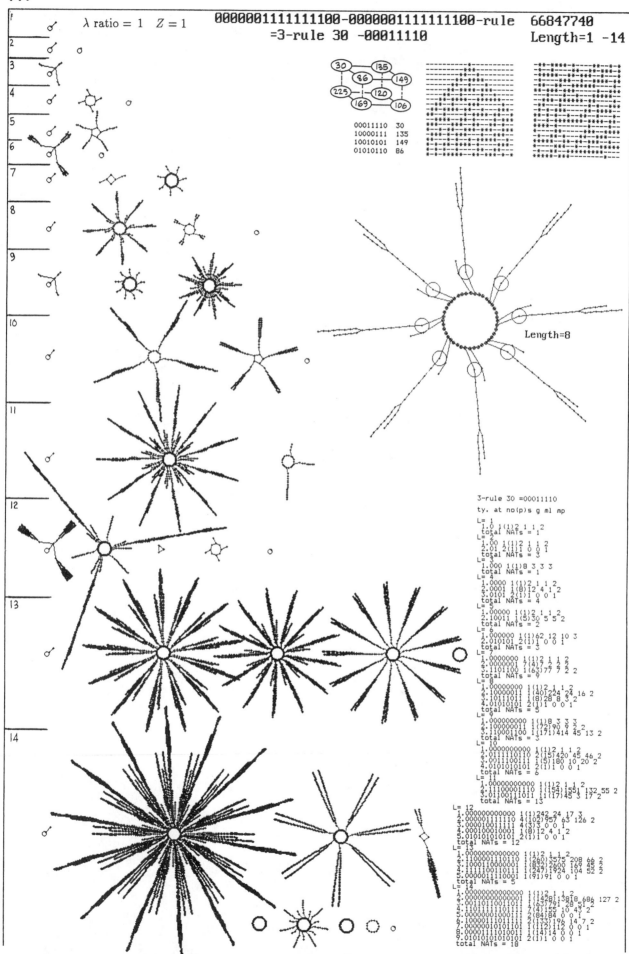

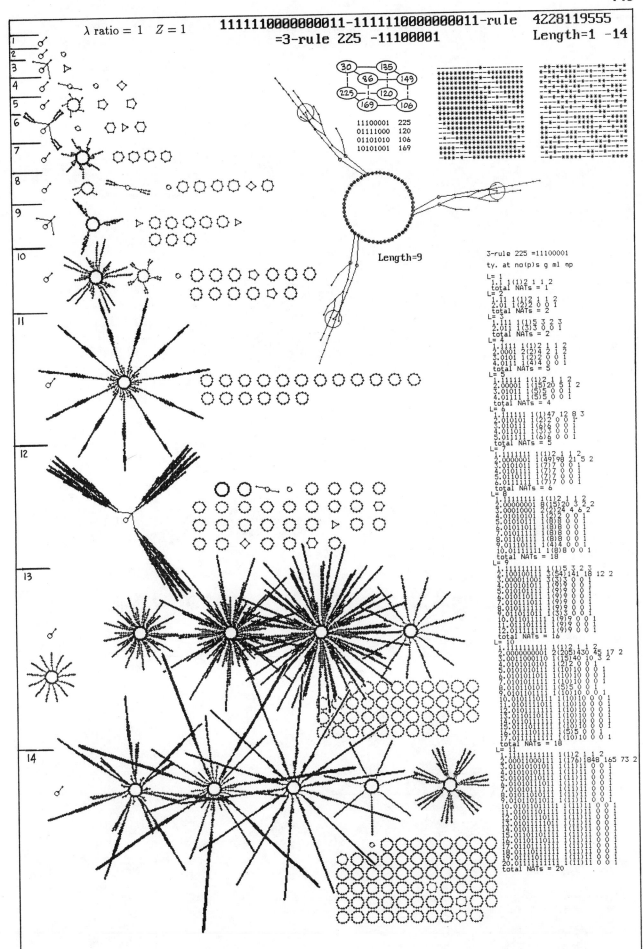

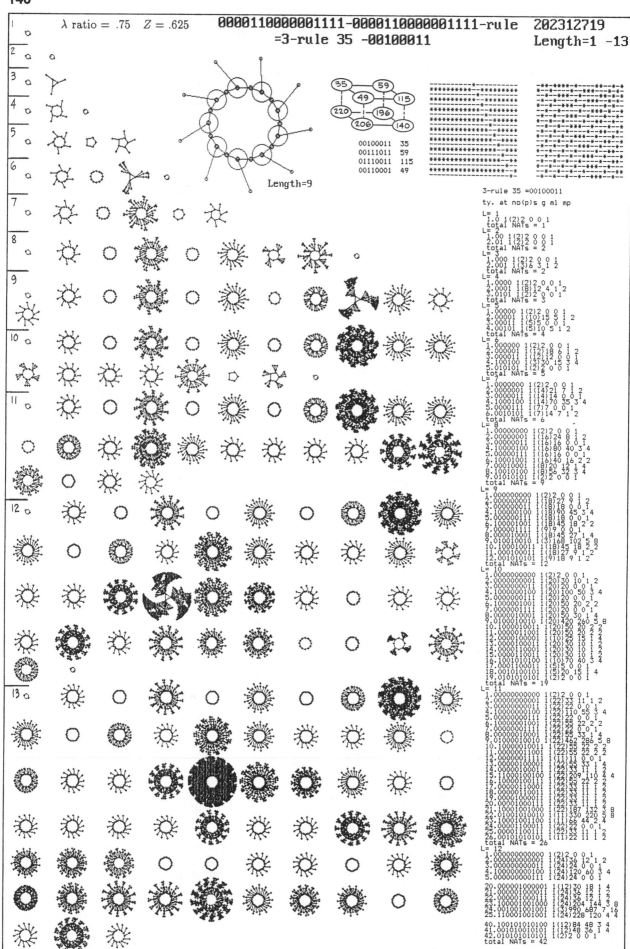

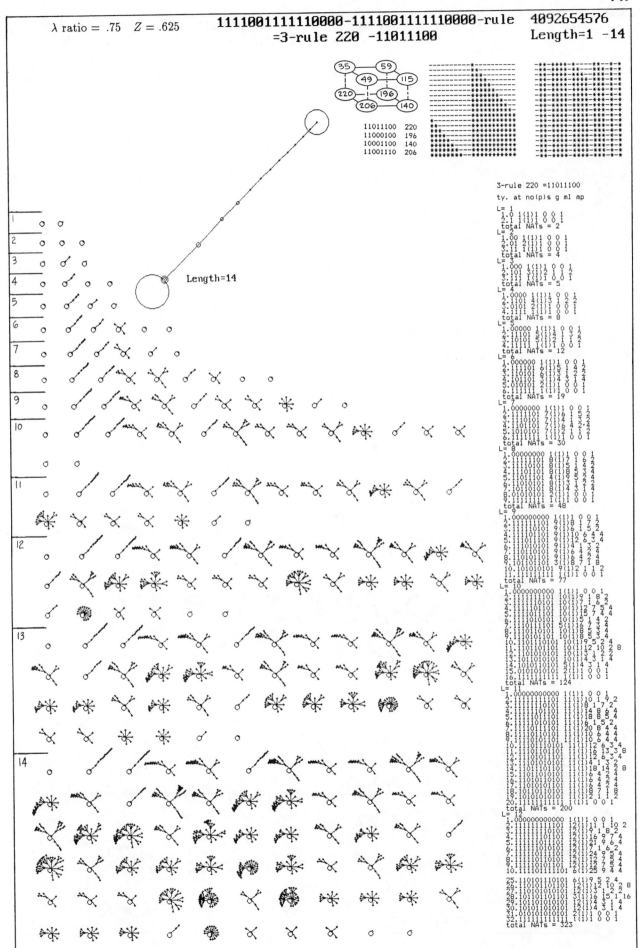

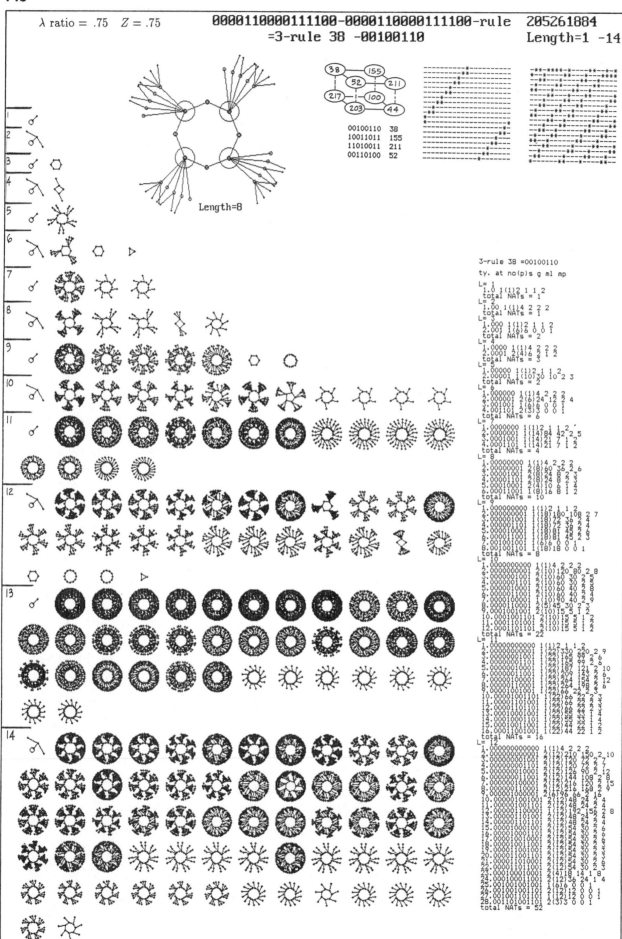

λ ratio = .75   Z = .75   **1111001111000011-1111001111000011-rule   4089705411**
                          =3-rule 217 -11011001   Length=1 -15

```
11011001  217
01100100  100
00101100   44
11001011  203
```

Length=10

```
3-rule 217 =11011001
ty. at no(p)s g ml mp
L= 1
1.1      1(1)2  1 1 2
 total NATs = 1
L= 2
1.11     1(1)4  2 2 2
 total NATs = 1
L= 3
1.111    1(1)2  1 1 2
2.001    1(3)3  0 0 1
3.011    3(1)1  0 0 1
 total NATs = 5
L= 4
1.1111   1(1)4  2 2 2
2.1110   4(1)3  1 2 2
 total NATs = 5
L= 5
1.11111  1(1)2  1 1 2
2.11110  5(1)6  2 3 3
 total NATs = 6
L= 6
1.111111   1(1)4  2 2 2
2.111110   6(1)9  2 3 4
3.001001   1(3)3  0 0 1
4.011011   3(1)1  0 0 1
 total NATs = 11
L= 7
1.1111111  1(1)2  1 1 2
2.1111110  7(1)15 7 4 5
3.1101110  7(1)3  1 2 2
 total NATs = 15
L= 8
1.11111111 1(1)4  2 2 2
2.11111110 8(1)21 7 4 6
3.11011110 8(1)7  3 3 3
4.11011110 8(1)7  3 3 3
 total NATs = 21
L= 9
1.111111111 1(1)2  1 1 2
2.111111110 9(1)28 16 4 7
3.111011110 9(1)14 7 3 6
4.111011011 9(1)14 7 4 4
5.001001001 1(3)3  0 0 1
6.011011011 3(1)1  0 0 1
 total NATs = 32
L= 10
1.1111111111 1(1)4  2 2 2
2.1111111110 10(1)36 22 5 8
3.1111101110 10(1)24 13 3 8
4.1111011110 10(1)28 16 3 9
5.1111011011 10(1)25 14 3 5
6.1101101110 10(1)3  1 2 2
 total NATs = 46
L= 11
1.11111111111 1(1)2  1 1 2
2.11111111110 11(1)45 29 4 9
3.11111101110 11(1)38 22 4 10
4.11110111110 11(1)40 30 5 12
5.11110110110 11(1)40 24 6 6
6.11101101110 11(1)25 14 3 5
7.11011011011 11(1)7  3 3 3
 total NATs = 67
L= 12
1.111111111111 1(1)4  2 2 2
2.111111111110 12(1)55 37 5 10
3.111111101110 12(1)55 34 4 15
4.111110111110 12(1)76 50 4 7
5.111110110110 12(1)58 38 6 7
6.111011011110 6(1)83 54 3 16
7.111011011010 12(1)18 10 3 6
8.111011011010 4(1)8 10 2 8
9.110110110110 12(1)17 10 3 6
10.110110110110 12(1)17 10 3 6
11.001001001001 12(1)14 7 4 4
12.010110110110 1(3)1 0 0 1
 total NATs = 99
L= 13
1.1111111111111 1(1)2 1 1 2
2.1111111111110 13(1)4 4 4 4
3.1111111101110 13(1)98 46 4 11
4.1111110111110 13(1)110 77 5 18
5.1111101111110 13(1)82 56 7 8
6.1111011011110 13(1)129 88 4 20
7.1111011011010 13(1)135 21 4 8
8.1111011010110 13(1)38 24 4 2
9.1110110110110 13(1)25 14 2 6
10.1101101101110 13(1)31 20 3 8
11.1101101101011 13(1)3 1 2 2
 total NATs = 144
L= 14
1.11111111111111 1(1)4 2 2 2
2.11111111111110 14(1)178 59 5 12
3.11111111101110 14(1)98 67 4 16
4.11111110111110 14(1)191 5 21
5.11111101111110 14(1)109 78 4 9
6.11110110110110 14(1)186 133 5 24
7.11110110110110 7 14(1)199 141 4 25
8.11110110110110 14(1)160 30 5 10
9.11101101101110 14(1)63 41 3 16
10.11101101101010 14(1)73 49 4 12
11.11101101101011 14(1)141 25 4 6
12.11011011011010 14(1)172 40 6 18
13.11011011011011 14(1)15 34 4 24
14.11011011011011 14(1)69 47 3 12
15.11011011010110 14(1)29 4 2 4
16.01011011011011 1(3)3 0 0 1
17.11011011011110 14(1)7 3 3 3
 total NATs = 211
L= 15
1.111111111111111 1(1)2 1 1 2
2.111111111111110 15(1)191 67 4 13
3.111111111101110 15(1)124 88 5 18
4.111111110111110 15(1)199 152 5 24
5.111111011111110 15(1)140 104 7 10
6.111110110110110 15(1)204 189 6 7
7.111101101101110 15(1)184 210 6 20
8.111011011011110 15(1)64 63 6 12
9.111011011010110 15(1)100 68 4 20
10.111110110110110 15(1)123 86 5 15
20.110110110110110 15(1)17 10 3 6
21.110110110110111 15(1)14 7 4 3
22.110110110110111 15(1)21 13 3 6
23.001001001001001 1(3)3 0 0 1
24.010110110110110 3(1)1 0 0 1
 total NATs = 310
```

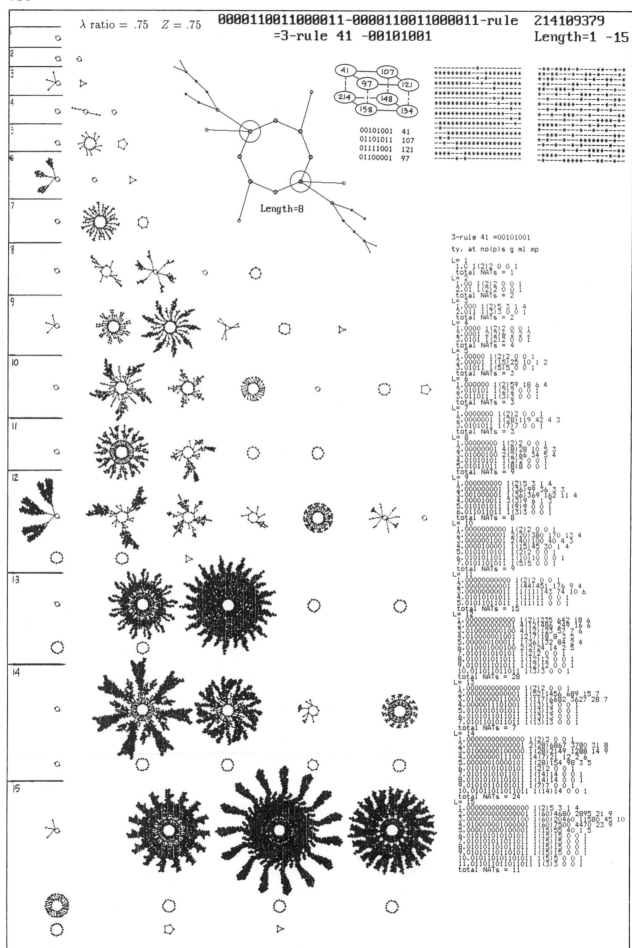

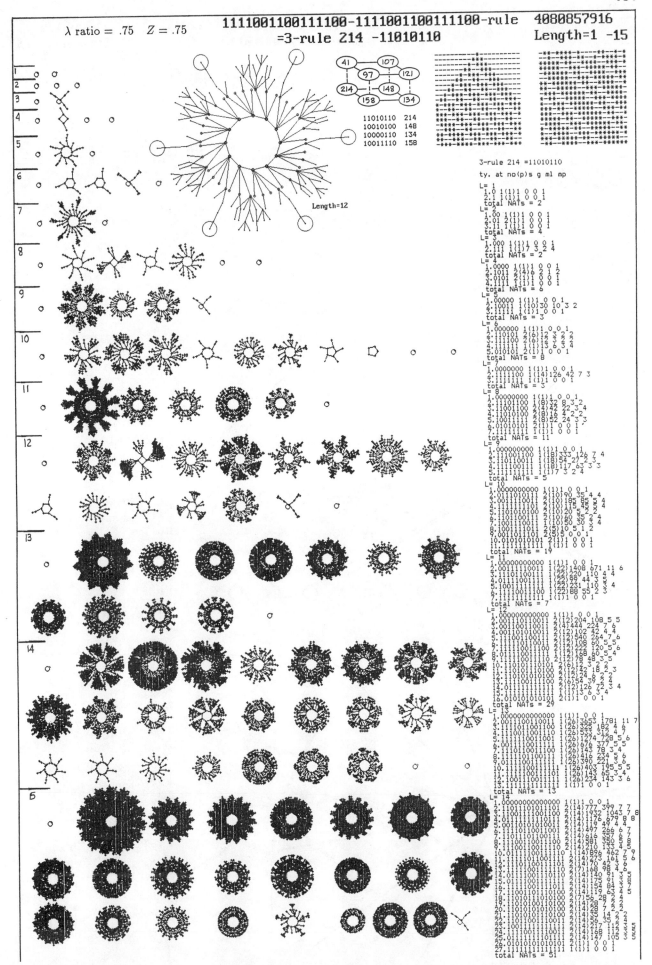

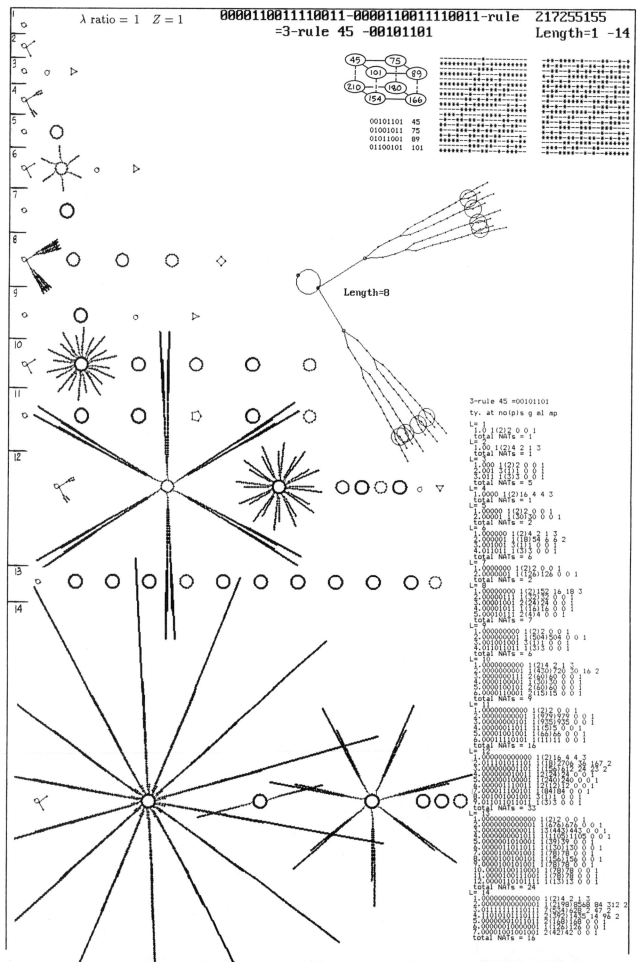

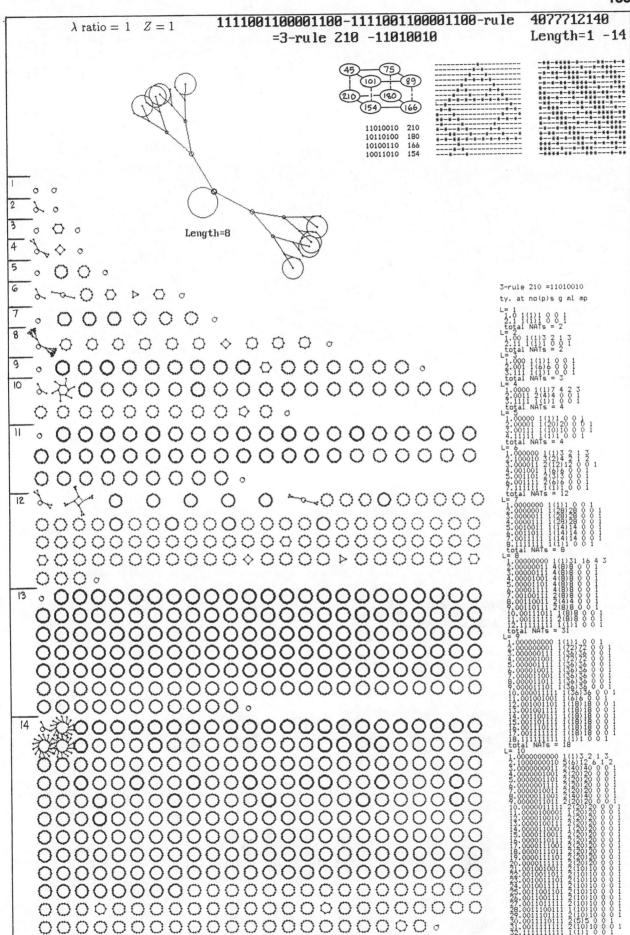

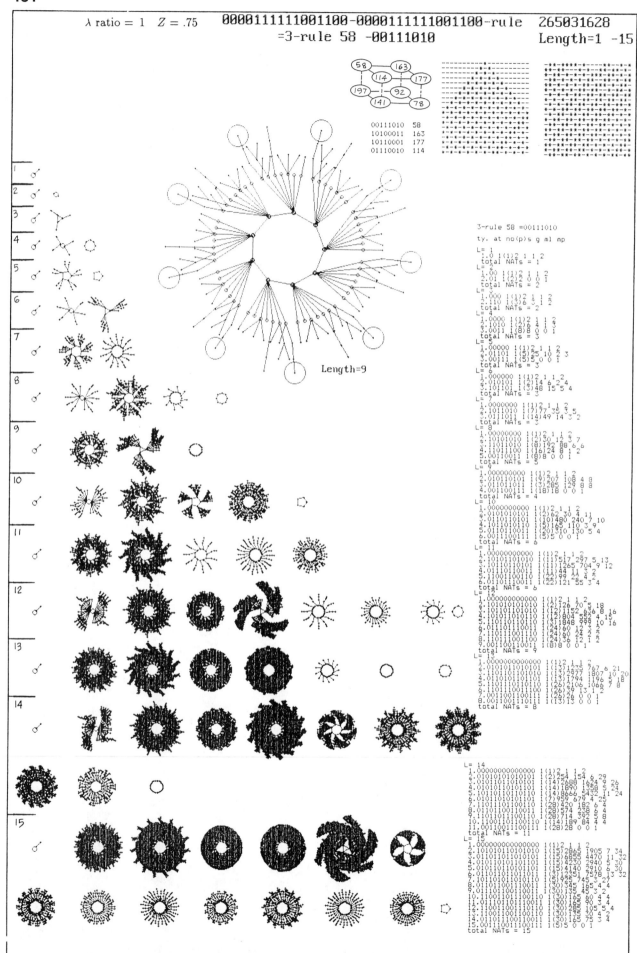

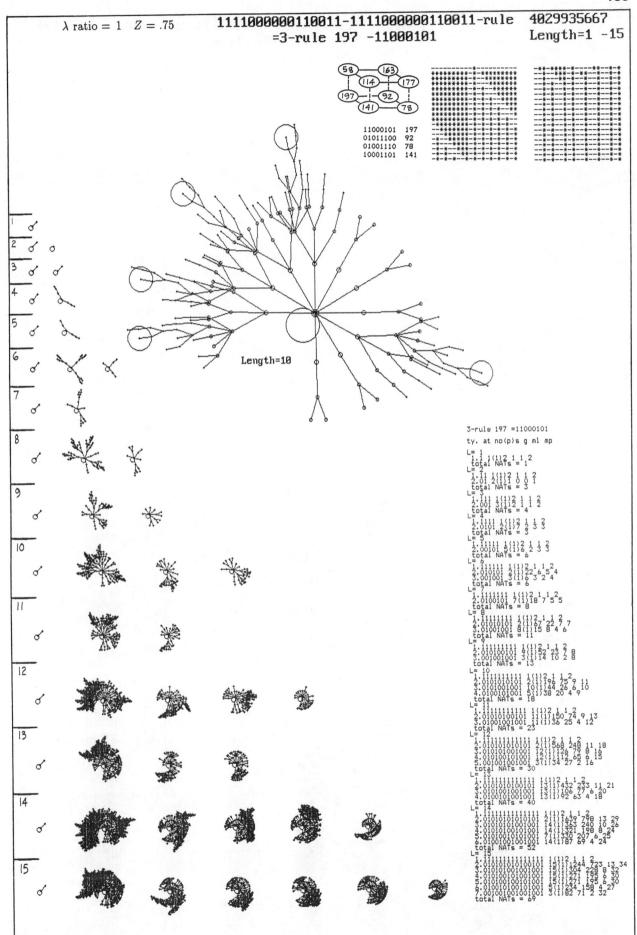

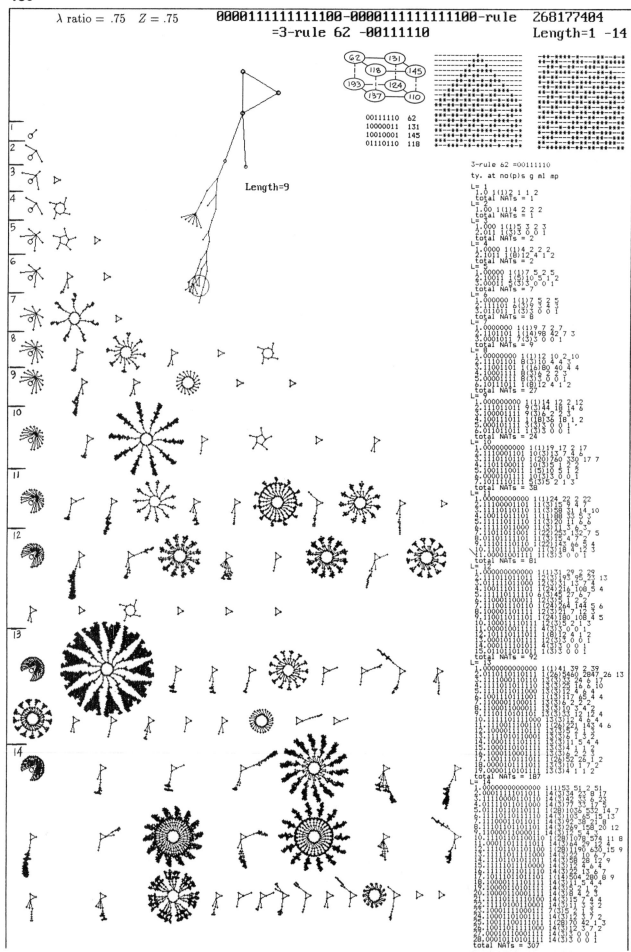

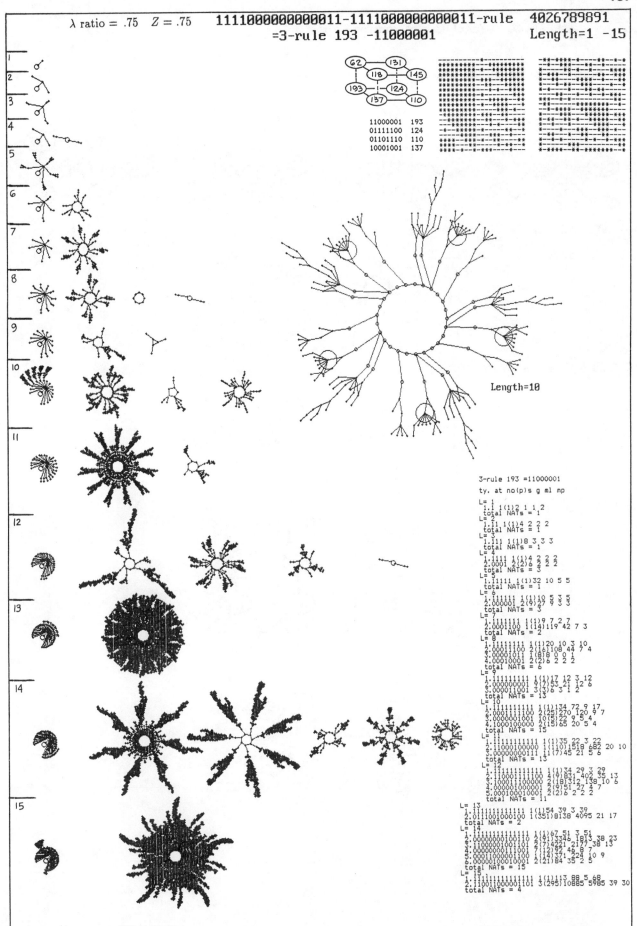

158 APPENDIX 2 Atlas of Basin of Attraction Fields

A2.3.4 Fully Asymmetric Rule Clusters (see section 3.3.8)

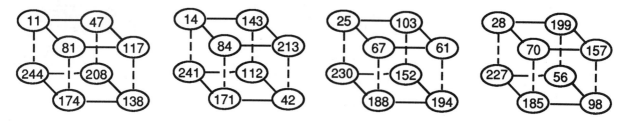

The cluster will collapse if, for a given rule R, $R_c = R_n$,

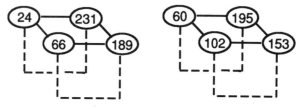

And if, for a given rule R, $R_c = R_{nr}$,

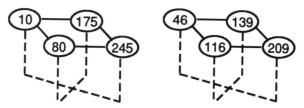

And if, for a given rule R, $R_n = R_r$,

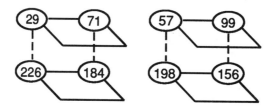

And also if $R = R_n$, then

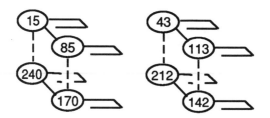

A2.3 n=3 Rules

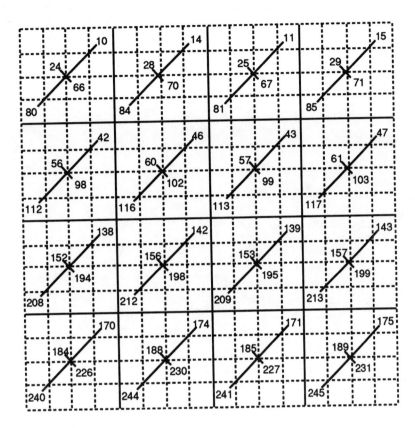

The rule-space matrix (see appendix 4).

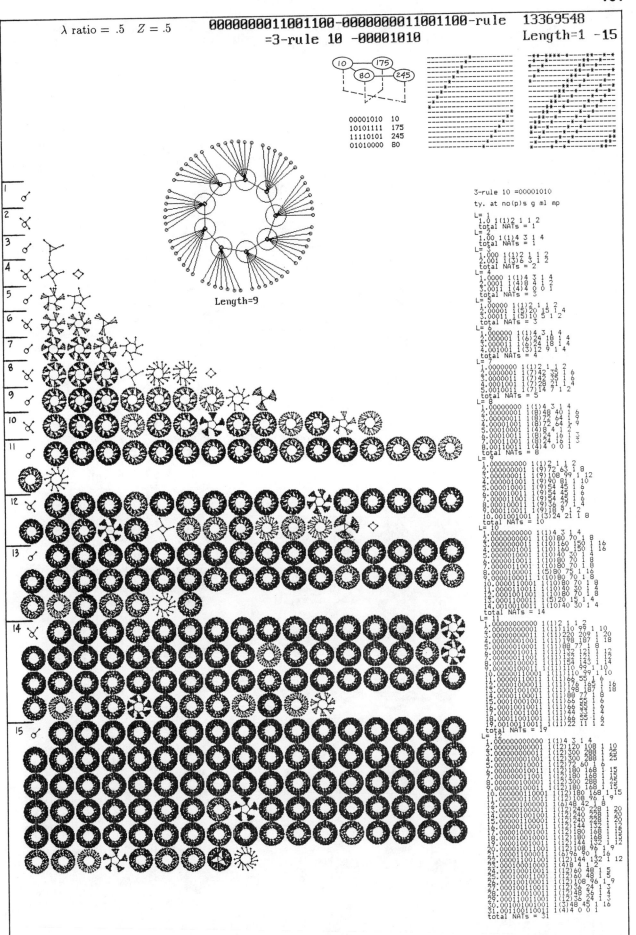

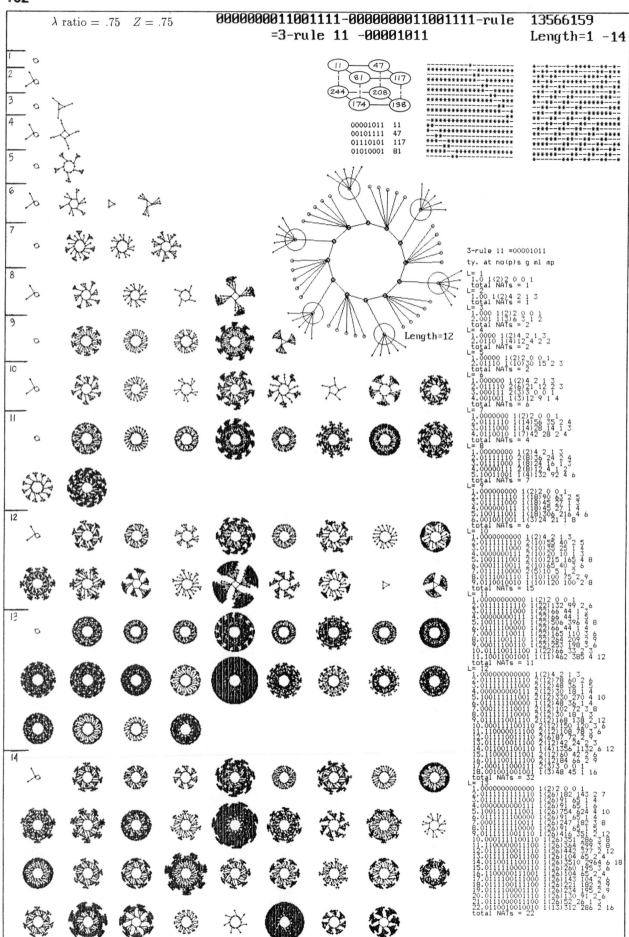

163

λ ratio = .75   Z = .75   1111111100110000-1111111100110000-rule   4281401136
                =3-rule 244 -11110100   Length=1 -13

```
11110100  244
11010000  208
10001010  138
10101110  174
```

Length=13

```
3-rule 244 =11110100
ty. at no(p)s g ml mp
L= 1
1.0   1(1)1 0 0 1
2.1   1(1)1 0 0 1
total NATs = 2
L= 2
1.00  1(1)1 0 0 1
2.11  1(1)3 2 1 3
total NATs = 2
L= 3
1.000  1(1)1 0 0 1
2.101  1(3)6 3 1 2
3.111  1(1)1 0 0 1
total NATs = 3
L= 4
1.0000 1(1)1 0 0 1
2.1001 1(4)8 4 1 2
3.1111 1(1)3 2 1 3
4.0111 1(4)4 2 0 1
total NATs = 4
L= 5
1.00000  1(1)1 0 0 1
2.10001  1(5)10 5 1 2
3.10111  1(5)15 10 1 3
4.00111  1(5)5 0 0 1
5.11111  1(1)1 0 0 1
total NATs = 5
L= 6
1.000000 1(1)1 0 0 1
2.100001 1(6)12 6 1 2
3.100111 1(6)18 12 1 3
4.000111 1(6)6 0 0 1
5.100101 1(3)12 9 1 4
6.111101 1(6)12 3 2 3
7.111011 1(6)12 6 1 2
total NATs = 7
L= 7
1.0000000 1(1)1 0 0 1
2.1000001 1(7)14 7 1 2
3.1000111 1(7)21 14 1 3
4.0000111 1(7)7 0 0 1
5.1001101 1(7)28 21 1 4
6.1000111 1(7)28 21 1 4
7.1011011 1(7)14 7 1 2
8.1111111 1(1)1 0 0 1
total NATs = 9
L= 8
1.00000000 1(1)1 0 0 1
2.10000001 1(8)16 8 1 2
3.10000111 1(8)24 16 1 3
4.00000111 1(8)8 0 0 1
5.10011011 1(8)32 24 1 4
6.10010011 1(4)16 12 1 4
7.10011101 1(8)32 24 1 4
8.10001111 1(8)16 8 1 2
9.10011111 1(8)16 8 1 2
10.10101011 1(8)48 40 1 6
11.10100011 1(8)16 8 1 2
12.11111111 1(1)3 2 1 3
13.01111111 1(8)24 16 1 3
total NATs = 14
L= 9
1.000000000 1(1)1 0 0 1
2.100000001 1(9)18 9 1 2
3.100000111 1(9)27 18 1 3
4.000000111 1(9)18 9 1 2
5.100001101 1(9)36 27 1 4
6.100011011 1(9)36 27 1 4
7.100011111 1(9)18 9 1 2
8.100101101 1(9)36 27 1 4
9.100110011 1(9)18 9 1 2
10.100110111 1(9)54 45 1 6
11.100101101 1(9)18 9 1 2
12.100111101 1(9)54 45 1 6
13.100111111 1(9)36 27 1 4
14.101010111 1(9)24 2 1 2
15.101011111 1(3)45 45 1 5
16.101101111 1(9)27 18 1 3
17.101011111 1(9)27 18 1 3
18.101111111 1(9)36 27 1 4
19.011111111 1(9)27 18 1 3
20.111111111 1(1)1 0 0 1
total NATs = 20
L= 10
1.0000000000 1(1)1 0 0 1
2.1000000001 1(10)20 10 1 2
3.1000000111 1(10)30 20 1 3
4.0000000111 1(10)20 10 1 2
5.1000011011 1(10)40 30 1 4
6.1000011101 1(10)40 30 1 4
7.1000111111 1(10)40 30 1 4
8.1000011111 1(10)20 10 1 2
9.1001000111 1(5)20 15 1 4
10.1000110001 1(5)20 15 1 4
11.1001100111 1(10)60 50 1 6
12.1001101011 1(10)60 50 1 6
13.1001110011 1(10)60 50 1 6
14.1001111001 1(10)60 50 1 6
15.1001011111 1(10)60 50 1 6
16.1001101101 1(10)80 70 1 8
17.1001010101 1(5)20 15 1 4
18.1010001111 1(10)80 70 1 8
19.1010011111 1(10)40 30 1 4
20.1010101111 1(10)40 30 1 4
21.1011101111 1(10)20 10 1 2
22.1000010101 1(10)10 0 0 1
23.1011101111 1(10)10 0 0 1
24.1010111111 1(10)40 30 1 4
25.1011110111 1(10)40 30 1 4
26.1011111111 1(5)45 40 1 9
27.1011110111 1(10)30 20 1 3
28.1011111111 1(5)10 5 0 1
29.0111111111 1(10)30 20 1 3
30.1111111111 1(10)40 30 1 4
31.1111111111 1(10)20 10 1 2
total NATs = 31
```

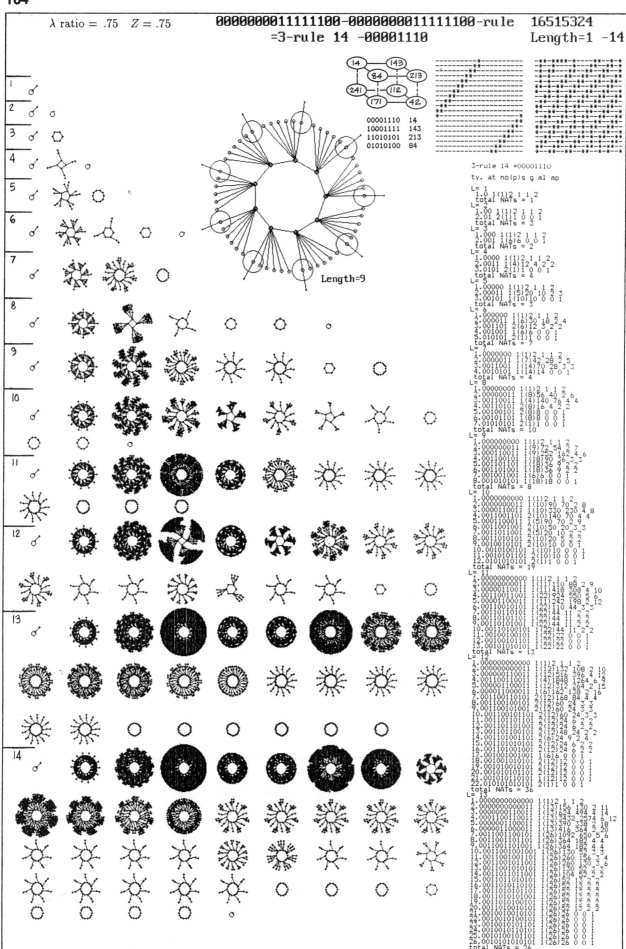

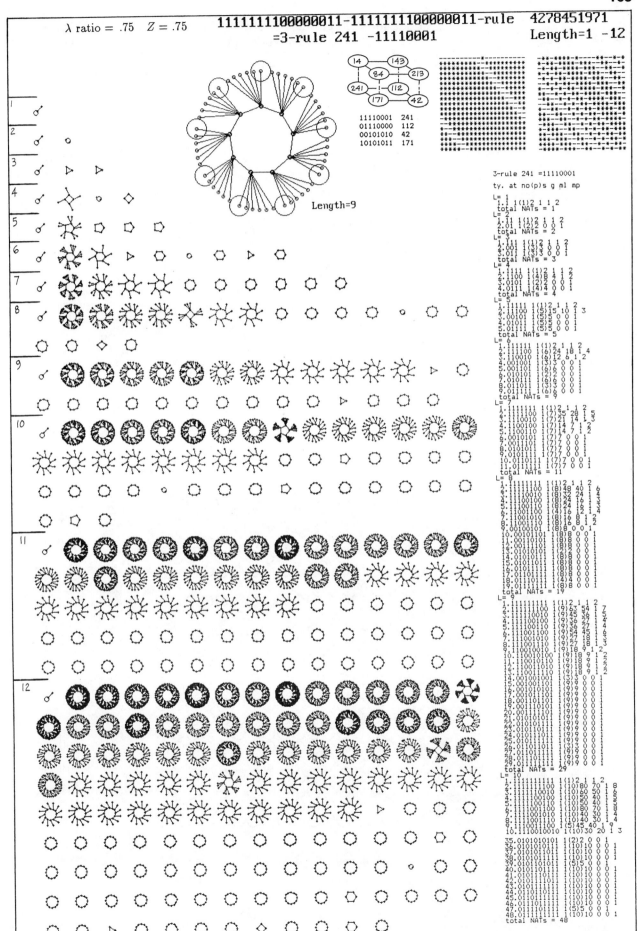

λ ratio = 1 Z = 1   0000000011111111-0000000011111111-rule   16711935
=3-rule 15 -00001111   Length=1 -13

```
00001111  15
01010101  85
```

Length=9

```
3-rule 15 =00001111
ty. at no(p)s g ml mp
L= 1
 1.0      1(2)2  0 0 1
  total NATs = 1
L= 2
 1.00     1(2)2  0 0 1
 2.01     2(1)1  0 0 1
  total NATs = 3
L= 3
 1.000    1(2)2  0 0 1
 2.001    1(6)6  0 0 1
  total NATs = 2
L= 4
 1.0000   1(2)2  0 0 1
 2.0001   2(4)4  0 0 1
 3.0011   1(4)4  0 0 1
 4.0101   2(1)1  0 0 1
  total NATs = 6
L= 5
 1.00000  1(2)2  0 0 1
 2.00001  1(10)10 0 0 1
 3.00011  1(10)10 0 0 1
 4.00101  1(10)10 0 0 1
  total NATs = 4
L= 6
 1.000000 1(2)2  0 0 1
 2.000001 2(6)6  0 0 1
 3.000011 2(6)6  0 0 1
 4.000101 2(6)6  0 0 1
 5.000111 2(3)3  0 0 1
 6.001001 1(6)6  0 0 1
 7.001011 2(6)6  0 0 1
 8.010101 2(1)1  0 0 1
  total NATs = 14
L= 7
 1.0000000 1(2)2  0 0 1
 2.0000001 1(14)14 0 0 1
 3.0000011 1(14)14 0 0 1
 4.0000101 1(14)14 0 0 1
 5.0000111 1(14)14 0 0 1
 6.0001001 1(14)14 0 0 1
 7.0001011 1(14)14 0 0 1
 8.0001101 1(14)14 0 0 1
 9.0010011 1(14)14 0 0 1
10.0010101 1(14)14 0 0 1
  total NATs = 10
L= 8
 1.00000000 1(2)2  0 0 1
 2.00000001 2(8)8  0 0 1
 3.00000011 2(8)8  0 0 1
 4.00000101 2(8)8  0 0 1
 5.00000111 2(8)8  0 0 1
 6.00001001 2(8)8  0 0 1
 7.00001011 2(8)8  0 0 1
 8.00001101 2(8)8  0 0 1
 9.00001111 1(8)8  0 0 1
10.00010001 2(4)4  0 0 1
11.00010011 2(8)8  0 0 1
12.00010101 2(8)8  0 0 1
13.00010111 2(8)8  0 0 1
14.00011001 2(8)8  0 0 1
15.00011011 2(8)8  0 0 1
16.00100101 2(8)8  0 0 1
17.00101011 2(8)8  0 0 1
18.00101101 1(8)8  0 0 1
19.00110101 1(4)4  0 0 1
20.01010101 2(1)1  0 0 1
  total NATs = 36
L= 9
 1.000000000 1(2)2  0 0 1
 2.000000001 1(18)18 0 0 1
 3.000000011 1(18)18 0 0 1
 4.000000101 1(18)18 0 0 1
 5.000000111 1(18)18 0 0 1
 6.000001001 1(18)18 0 0 1
 7.000001011 1(18)18 0 0 1
 8.000001101 1(18)18 0 0 1
 9.000001111 1(18)18 0 0 1
10.000010001 1(18)18 0 0 1
11.000010011 1(18)18 0 0 1
12.000010101 1(18)18 0 0 1
13.000010111 1(18)18 0 0 1
14.000011001 1(18)18 0 0 1
15.000011011 1(18)18 0 0 1
16.000011101 1(18)18 0 0 1
17.000100101 1(18)18 0 0 1
18.000100111 1(18)18 0 0 1
19.000101011 1(18)18 0 0 1
20.000101101 1(18)18 0 0 1
21.001010101 1(18)18 0 0 1
22.001010111 1(18)18 0 0 1
23.001011011 1(18)18 0 0 1
24.001011101 1(18)18 0 0 1
25.001100101 1(18)18 0 0 1
26.001001001 1(6)6  0 0 1
27.001001011 1(18)18 0 0 1
28.001001101 1(18)18 0 0 1
29.001010011 1(18)18 0 0 1
30.001010101 1(18)18 0 0 1
  total NATs = 30
```

λ ratio = 1   Z = 1   **1111111100000000-1111111100000000-rule   4278255360**
                     =3-rule 240 -11110000   Length=1 -12

```
11110000  240
10101010  170
```

Length=9

```
3-rule 240 =11110000
ty. at no(p)s g ml mp
L= 1
1.0  1(1)1  0 0 1
2.1  1(1)1  0 0 1
 total NATs = 2
L= 2
1.00  1(1)1  0 0 1
2.01  1(2)2  0 0 1
3.11  1(1)1  0 0 1
 total NATs = 3
L= 3
1.000  1(1)1  0 0 1
2.001  1(3)3  0 0 1
3.011  1(3)3  0 0 1
4.111  1(1)1  0 0 1
 total NATs = 4
L= 4
1.0000  1(1)1  0 0 1
2.0001  1(4)4  0 0 1
3.0011  1(4)4  0 0 1
4.0101  1(2)2  0 0 1
5.0111  1(4)4  0 0 1
6.1111  1(1)1  0 0 1
 total NATs = 6
L= 5
1.00000  1(1)1  0 0 1
2.00001  1(5)5  0 0 1
3.00011  1(5)5  0 0 1
4.00101  1(5)5  0 0 1
5.00111  1(5)5  0 0 1
6.01011  1(5)5  0 0 1
7.01111  1(5)5  0 0 1
8.11111  1(1)1  0 0 1
 total NATs = 8
L= 6
1.000000  1(1)1  0 0 1
2.000001  1(6)6  0 0 1
3.000011  1(6)6  0 0 1
4.000101  1(6)6  0 0 1
5.000111  1(6)6  0 0 1
6.001001  1(3)3  0 0 1
7.001011  1(6)6  0 0 1
8.001101  1(6)6  0 0 1
9.001111  1(6)6  0 0 1
10.010101  1(2)2  0 0 1
11.010111  1(6)6  0 0 1
12.011011  1(3)3  0 0 1
13.011111  1(6)6  0 0 1
14.111111  1(1)1  0 0 1
 total NATs = 14
L= 7
1.0000000  1(1)1  0 0 1
2.0000001  1(7)7  0 0 1
3.0000011  1(7)7  0 0 1
4.0000101  1(7)7  0 0 1
5.0000111  1(7)7  0 0 1
6.0001001  1(7)7  0 0 1
7.0001011  1(7)7  0 0 1
8.0001101  1(7)7  0 0 1
9.0001111  1(7)7  0 0 1
10.0010101  1(7)7  0 0 1
11.0010111  1(7)7  0 0 1
12.0011011  1(7)7  0 0 1
13.0011101  1(7)7  0 0 1
14.0011111  1(7)7  0 0 1
15.0101011  1(7)7  0 0 1
16.0101111  1(7)7  0 0 1
17.0110111  1(7)7  0 0 1
18.0110111  1(7)7  0 0 1
19.0111111  1(7)7  0 0 1
20.1111111  1(1)1  0 0 1
 total NATs = 20
L= 8
1.00000000  1(1)1  0 0 1
2.00000001  1(8)8  0 0 1
3.00000011  1(8)8  0 0 1
4.00000101  1(8)8  0 0 1
5.00000111  1(8)8  0 0 1
6.00001001  1(8)8  0 0 1
7.00001011  1(8)8  0 0 1
8.00001101  1(8)8  0 0 1
9.00001111  1(8)8  0 0 1
10.00010001  1(4)4  0 0 1
11.00010011  1(8)8  0 0 1
12.00010101  1(8)8  0 0 1
13.00010111  1(8)8  0 0 1
14.00011001  1(8)8  0 0 1
15.00011101  1(8)8  0 0 1
16.00011111  1(8)8  0 0 1
17.00100101  1(8)8  0 0 1
18.00100101  1(8)8  0 0 1
19.00101011  1(8)8  0 0 1
20.00101101  1(8)8  0 0 1
21.00101111  1(8)8  0 0 1
22.00110011  1(4)4  0 0 1
23.00110101  1(8)8  0 0 1
24.00110111  1(8)8  0 0 1
25.00111011  1(8)8  0 0 1
26.00111101  1(8)8  0 0 1
27.00111111  1(8)8  0 0 1
28.00111111  1(8)8  0 0 1
29.01010101  1(2)2  0 0 1
30.01010111  1(8)8  0 0 1
31.01011011  1(8)8  0 0 1
32.01011111  1(8)8  0 0 1
33.01101111  1(8)8  0 0 1
34.01110111  1(4)4  0 0 1
35.01111111  1(8)8  0 0 1
36.11111111  1(1)1  0 0 1
 total NATs = 36
```

169

λ ratio = .5   Z = .5   0000001111000000-0000001111000000-rule   62915520
                                    =3-rule 24 -00011000   Length=1 -15

```
00011000  24
11100111  231
10111101  189
01000010  66
```

Length=9

```
3-rule 24 =00011000
ty. at no(p)s g ml mp
L= 1
 1.0 1(1)2 1 1 2
 total NATs = 1
L= 2
 1.00 1(1)4 3 1 4
 total NATs = 1
L= 3
 1.000 1(1)2 1 1 2
 2.001 1(3)6 3 1 2
 total NATs = 2
L= 4
 1.0000 1(1)8 5 2 4
 2.0001 1(4)8 4 1 2
 total NATs = 2
L= 5
 1.00000 1(1)2 1 1 2
 2.00001 1(5)30 20 2 4
 total NATs = 2
L= 6
 1.000000 1(1)4 3 1 4
 2.000001 1(6)48 36 2 4
 3.100100 1(3)12 9 1 4
 total NATs = 3
L= 7
 1.0000000 1(1)2 1 1 2
 2.0000001 1(7)84 35 1 6
 3.1000100 1(7)42 35 1 6
 total NATs = 3
L= 8
 1.00000000 1(1)8 5 2 4
 2.00000001 1(8)112 88 2 6
 3.10000100 1(8)96 80 2 8
 4.10001000 1(4)40 36 1 10
 total NATs = 4
L= 9
 1.000000000 1(1)2 1 1 2
 2.000000001 1(9)180 144 2 8
 3.100000100 1(9)144 126 2 10
 4.100001000 1(9)162 144 2 12
 5.100100100 1(3)24 21 1 8
 total NATs = 5
L= 10
 1.0000000000 1(1)4 3 1 4
 2.0000000001 1(10)240 200 2 8
 3.1000000100 1(10)240 210 2 12
 4.1000001000 1(10)240 220 2 16
 5.0100001000 1(5)180 160 2 16
 6.1001001000 1(10)120 110 1 12
 total NATs = 6
L= 11
 1.00000000000 1(1)2 1 1 2
 2.00000000001 1(11)330 275 2 10
 3.10000000100 1(11)330 297 2 14
 4.10000001000 1(11)396 363 2 18
 5.01000001000 1(11)528 484 2 20
 6.10010010000 1(11)264 242 2 20
 7.10001001000 1(11)198 187 1 18
 total NATs = 7
L= 12
 1.000000000000 1(1)8 5 2 4
 2.000000000001 1(12)408 348 2 10
 3.100000000100 1(12)480 432 2 16
 4.100000001000 1(12)552 516 2 22
 5.010000001000 1(12)864 792 2 24
 6.100010010000 1(6)384 360 2 28
 7.100001001000 1(12)456 430 2 20
 8.010010010000 1(12)432 408 2 24
 9.100100100100 1(12)432 408 2 24
 10.100010001000 1(4)104 100 1 24
 11.100100010000 1(3)48 45 1 16
 total NATs = 11
L= 13
 1.0000000000000 1(1)2 1 1 2
 2.0000000000001 1(13)546 468 2 12
 3.1000000000100 1(13)624 572 2 18
 4.1000000001000 1(13)780 728 2 24
 5.0100000001000 1(13)1170 1092 2 28
 6.1000010010000 1(13)1248 1170 2 30
 7.1000001001000 1(13)624 585 2 24
 8.0100010010000 1(13)624 598 2 30
 9.1001001001000 1(13)624 598 2 30
 10.1000100010000 1(13)702 674 2 32
 11.1001000100000 1(13)546 520 2 24
 12.1001001001000 1(13)312 299 1 24
 total NATs = 12
L= 14
 1.00000000000000 1(1)4 3 1 4
 2.00000000000001 1(14)672 588 2 12
 3.10000000000100 1(14)840 770 2 20
 4.10000000001000 1(14)1008 952 2 28
 5.01000000001000 1(14)1680 1568 2 32
 6.10000001001000 1(14)1680 1588 2 36
 7.01000010010000 1(14)840 798 2 28
 8.10010010010000 1(14)1008 966 2 36
 9.10001001001000 1(7)1008 945 2 36
 10.10001001001000 1(14)1008 966 2 36
 11.10010010001000 1(14)1344 1288 2 40
 12.10010001000100 1(14)1344 1288 2 40
 13.01001001000100 1(14)1512 1456 2 48
 14.10010001000100 1(14)672 644 2 32
 15.10010010010010 1(14)504 490 1 32
 16.10010010010010 1(14)504 490 1 32
 17.10010010010010 1(7)252 245 1 36
 total NATs = 17
L= 15
 1.000000000000000 1(1)2 1 1 2
 2.000000000000001 1(15)840 735 2 14
 3.100000000000100 1(15)1050 975 2 22
 4.100000000001000 1(15)1350 1275 2 30
 5.010000000001000 1(15)2040 1920 2 36
 6.100000001001000 1(15)2400 2280 2 40
 7.010000010010000 1(15)1200 1140 2 32
 8.100100100100000 1(15)1350 1305 2 42
 9.100100010001000 1(5)450 435 2 42
 10.100010010010000 1(15)1620 1575 2 54
 11.010010010010000 1(15)1620 2070 2 48
 12.010010010010100 1(15)2160 2100 2 48
 13.010010010001100 1(15)2160 2100 2 48
 14.010010010000100 1(15)2160 2100 2 48
 15.010010010000000 1(15)1920 1860 2 60
 16.001001001001000 1(15)2160 2100 2 60
 17.100100100100000 1(15)1080 1050 2 48
 18.100010001000100 1(15)1080 1040 2 64
 19.010010001000100 1(15)1080 1050 2 48
 20.100010010010100 1(15)1080 1050 2 48
 21.100010010010100 1(15)1080 1050 2 48
 22.100010010010100 1(15)810 795 1 54
 23.100100100100100 1(3)96 93 1 32
 total NATs = 23
```

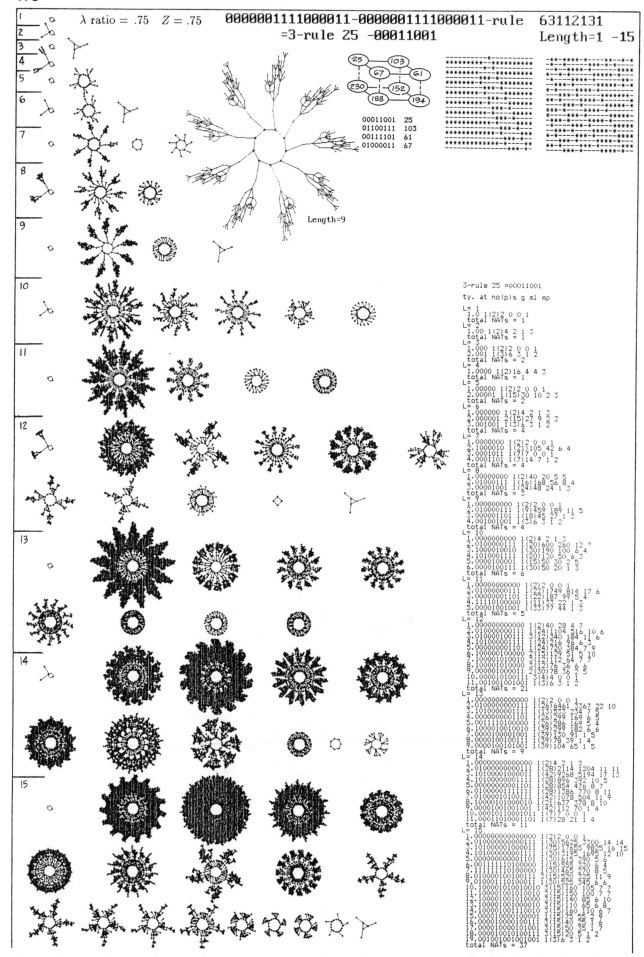

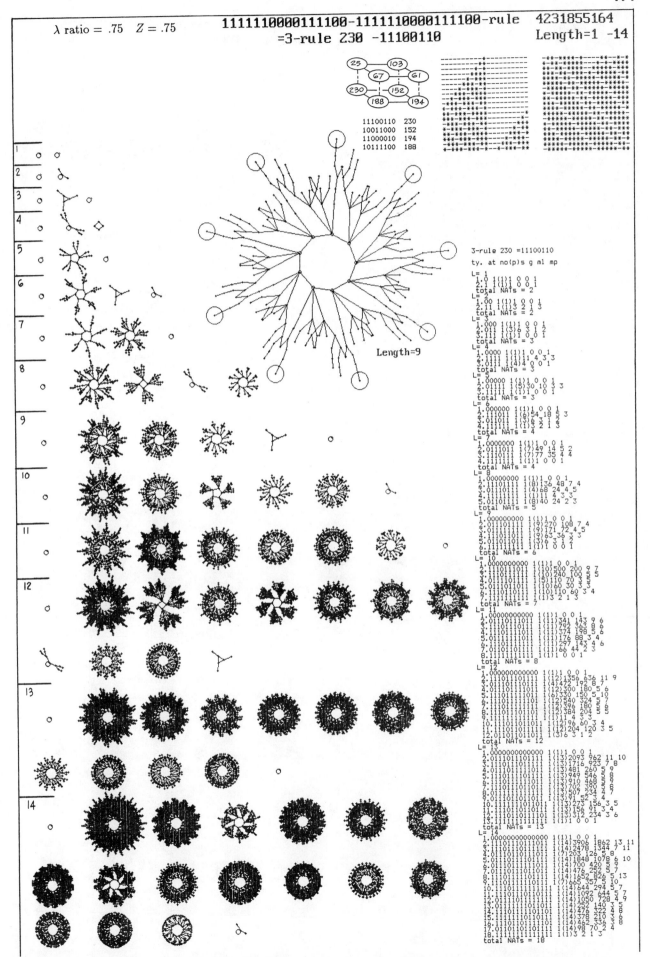

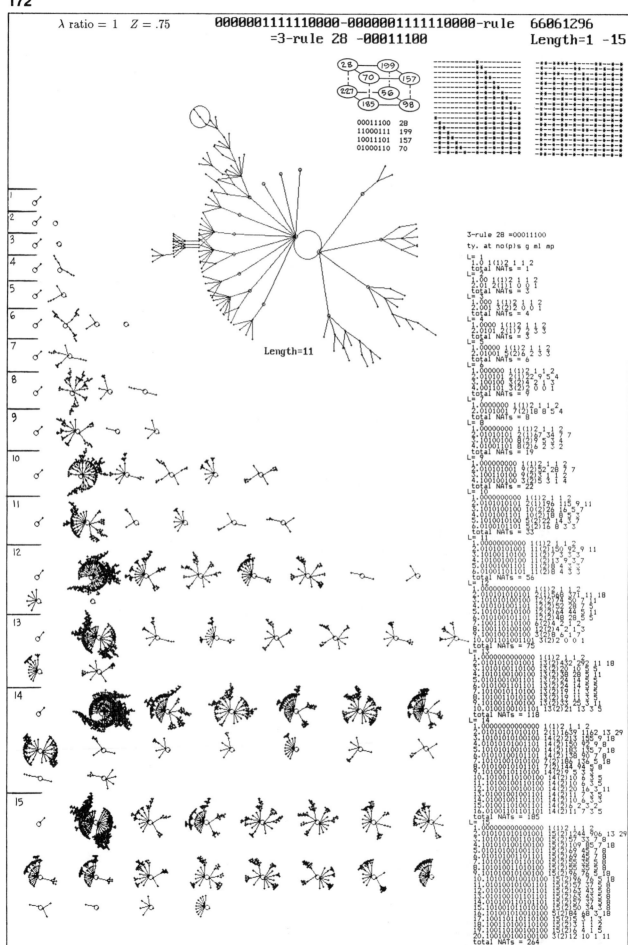

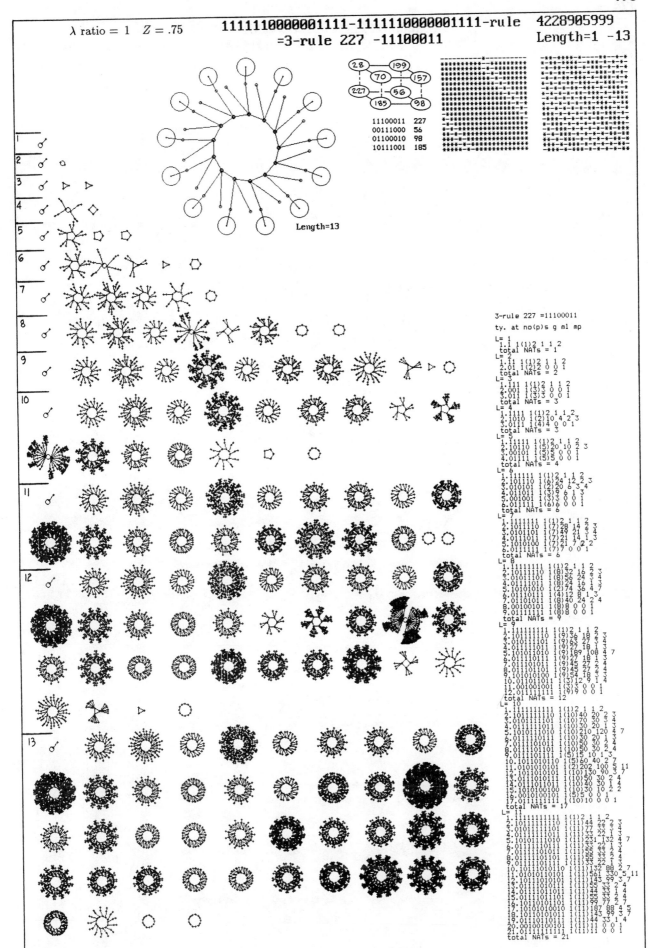

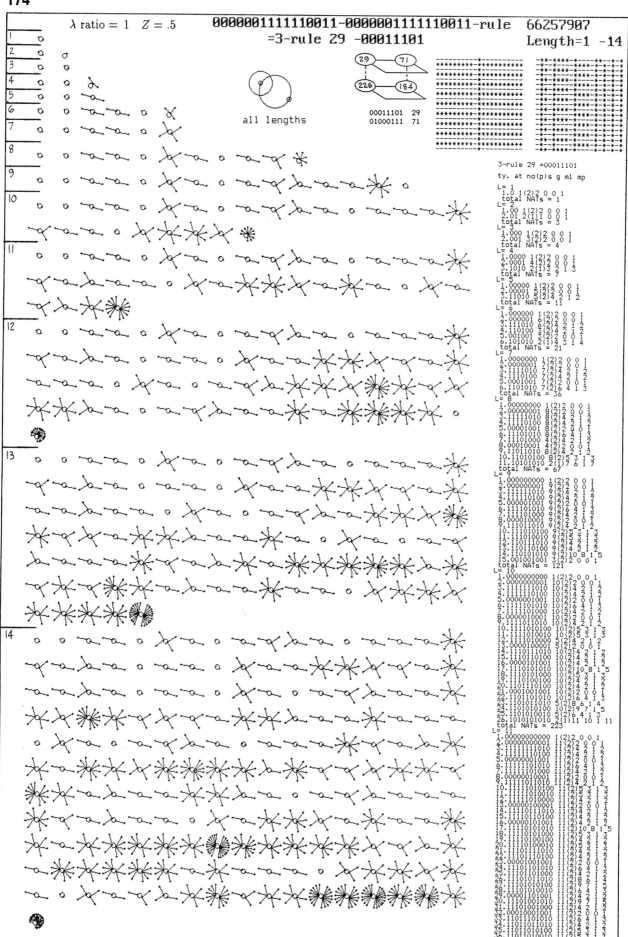

175

λ ratio = 1   Z = .5   1111110000001100-1111110000001100-rule   4228709388
                      =3-rule 226 -11100010   Length=1 -12

```
11100010  226
10111000  184
```

Length=9

3-rule 226 =11100010
ty. at no(p)s g ml mp

L= 1
1.0   1((1)1 0 0 1
2.1   1((1)1 0 0 1
total NATs = 2

L= 2
1.00 1((1)1 0 0 1
2.01 1((2)2 0 0 1
3.11 1((1)1 0 0 1
total NATs = 3

L= 3
1.000 1((1)1 0 0 1
2.001 1((3)3 0 0 1
4.011 1((3)3 0 0 1
4.111 1((1)1 0 0 1
total NATs = 4

L= 4
1.0000   1((1)1 0 0 1
2.0001   1((4)4 0 0 1
3.0101   1((2)6 4 1 3
4.0111   1((4)4 0 0 1
5.1111   1((1)1 0 0 1
total NATs = 5

L= 5
1.00000   1((1)1 0 0 1
2.00001   1((5)5 0 0 1
4.00101   1((5)10 5 1 2
4.01011   1((5)10 5 1 2
5.01111   1((5)5 0 0 1
6.11111   1((1)1 0 0 1
total NATs = 6

L= 6
1.000000   1((1)1 0 0 1
2.000001   1((6)6 0 0 1
3.000101   1((6)12 6 1 2
4.010101   1((6)20 12 2 4
5.001101   1((3)3 0 0 1
6.010111   1((6)12 6 1 2
7.011011   1((3)3 0 0 1
8.011111   1((6)6 0 0 1
9.111111   1((1)1 0 0 4
total NATs = 9

L= 7
1.0000000   1((1)1 0 0 1
2.0000001   1((7)7 0 0 1
3.0000101   1((7)14 7 1 2
4.0010101   1((7)35 21 2 3
5.0001001   1((7)7 0 0 1
6.0101011   1((7)35 21 2 3
7.0101111   1((7)14 7 1 2
8.0111111   1((7)7 0 0 1
9.1111111   1((1)1 0 0 1
total NATs = 10

L= 8
1.00000000   1((1)1 0 0 1
2.00000001   1((8)8 0 0 1
3.00000101   1((8)16 8 1 2
4.00010101   1((8)40 24 2 3
5.00001001   1((8)8 0 0 1
6.01010101   1((2)70 44 3 7
7.00010001   1((4)4 0 0 1
8.00101011   1((8)16 8 1 2
9.01010111   1((8)40 24 2 3
10.01011011 1((8)16 8 1 2
11.01010111 1((8)16 8 1 2
12.01101101 1((8)8 0 0 1
13.01101111 1((4)4 0 0 1
14.01111111 1((8)8 0 0 1
15.11111111 1((1)1 0 0 1
total NATs = 15

L= 9
1.000000000   1((9)9 0 0 1
2.000000001   1((9)9 0 0 1
3.000000101   1((9)18 9 1 2
4.000010101   1((9)45 27 2 3
5.000001001   1((9)9 0 0 1
6.010101011   1((9)9 0 0 1
7.000101011   1((9)126 81 3 5
8.000100101   1((9)18 9 1 2
9.001010111   1((9)18 9 1 2
10.001000101 1((9)18 9 1 2
11.001010101 1((9)45 27 2 3
12.010010101 1((3)33 0 0 1
13.010110111 1((9)18 9 1 2
14.010101111 1((9)18 9 1 2
15.010111111 1((9)18 9 1 2
16.011011011 1((9)9 0 0 1
17.011011111 1((9)9 0 0 1
18.011111111 1((9)9 0 0 1
19.011111111 1((9)9 0 0 1
20.111111111 1((1)1 0 0 1
total NATs = 20

L= 10
1.0000000000   1((1)1 0 0 1
2.0000000001   1((10)20 10 1 2
3.0000000101   1((10)50 30 2 3
4.0000010101   1((10)10 0 0 1
5.0000100101   1((10)10 0 0 1
6.0001001001   1((10)10 0 0 1
7.0001001001   1((10)140 90 3 11
8.0010010101   1((10)20 10 1 2
9.0101010101   1((2)252 170 4 11
10.0010010101 1((5)5 0 0 1
11.0010001101 1((10)50 30 2 3
12.0010010111 1((10)20 10 1 2
13.0001010111 1((10)140 90 3 11
14.0010010101 1((10)10 0 0 1
15.0101010101 1((5)20 15 1 4
16.0010110111 1((10)50 30 2 3
17.0101010111 1((10)20 15 1 4
18.0101011011 1((10)20 15 1 4
19.0101110111 1((10)20 10 1 2
20.0101010111 1((10)20 10 1 2
21.0101101111 1((10)20 10 1 2
22.0110110111 1((10)20 10 1 2
23.0110111111 1((10)20 10 1 2
24.0101011111 1((10)10 0 0 1
25.0111011111 1((10)10 0 0 1
26.0111101111 1((5)5 0 0 1
27.0111111111 1((10)10 0 0 1
28.0111111111 1((10)10 0 0 1
29.1111111111 1((1)1 0 0 1
total NATs = 29
```
```

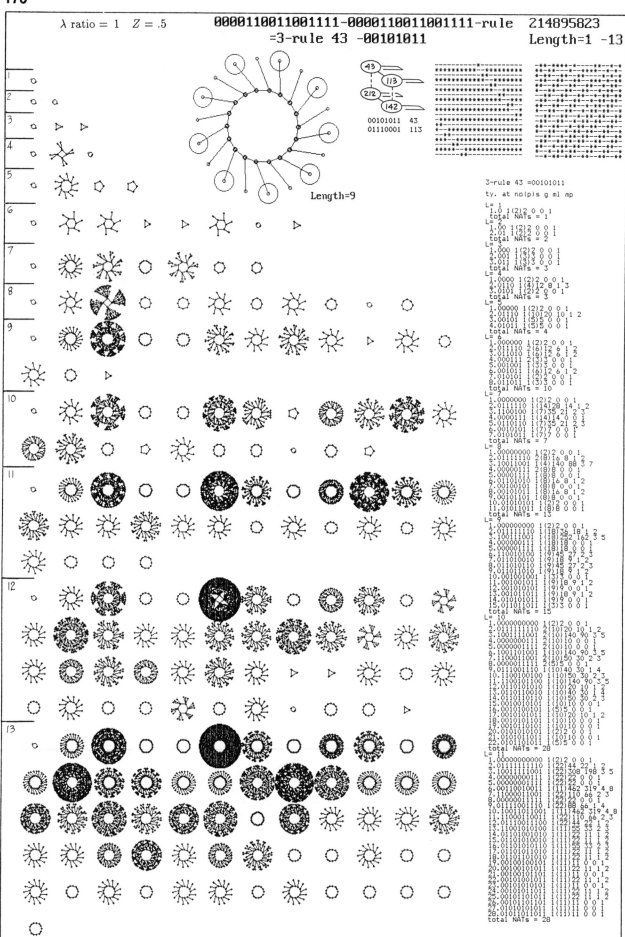

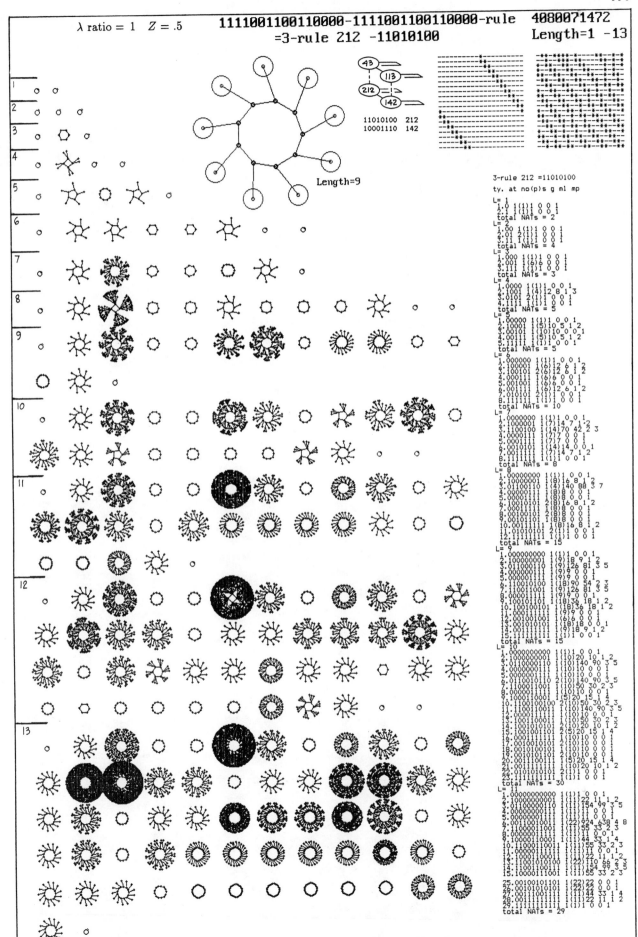

179

λ ratio = 1   Z = .5   **0000110011111100-0000110011111100-rule**   217844988
                        =3-rule 46 -00101110   Length=1 -15

```
00101110  46
10001011  139
11010001  209
01110100  116
```

Length=8

```
3-rule 46 =00101110
ty. at no(p)s g ml mp
L= 1
 1.0  1(1)2 1 1 2
  total NATs = 1
L= 2
 1.00 1(1)4 2 2 2
  total NATs = 1
L= 3
 1.000 1(1)2 1 1 2
 2.011 1(3)6 3 1 2
  total NATs = 2
L= 4
 1.0000 1(1)4 2 2 2
 2.0011 1(4)12 8 1 3
  total NATs = 2
L= 5
 1.00000 1(1)2 1 1 2
 2.00011 1(5)30 20 2 4
  total NATs = 2
L= 6
 1.000000 1(1)4 2 2 2
 2.000011 1(6)48 36 2 5
 3.011011 1(3)12 9 1 4
  total NATs = 3
L= 7
 1.0000000 1(1)2 1 1 2
 2.0000011 1(7)84 63 2 6
 3.0011011 1(7)42 35 1 6
  total NATs = 3
L= 8
 1.00000000 1(1)4 2 2 2
 2.00000011 1(8)120 96 2 7
 3.00011011 1(8)96 80 2 8
 4.00110011 1(4)36 32 1 9
  total NATs = 4
L= 9
 1.000000000 1(1)2 1 1 2
 2.000000011 1(9)180 144 2 8
 3.000011011 1(9)144 126 2 10
 4.000110011 1(9)162 144 2 12
 5.011011011 1(3)24 21 1 8
  total NATs = 5
L= 10
 1.0000000000 1(1)4 2 2 2
 2.0000000011 1(10)240 200 2 9
 3.0000011011 1(10)240 210 2 12
 4.0000110011 1(10)240 220 2 15
 5.0001100011 1(5)180 160 2 16
 6.0011011011 1(10)120 110 1 12
  total NATs = 6
L= 11
 1.00000000000 1(1)2 1 1 2
 2.00000000011 1(11)330 275 2 10
 3.00000011011 1(11)330 297 2 14
 4.00000110011 1(11)396 363 2 18
 5.00001100011 1(11)528 484 2 20
 6.00011011011 1(11)264 242 2 16
 7.00110011011 1(11)198 187 1 18
  total NATs = 7
L= 12
 1.000000000000 1(1)4 2 2 2
 2.000000000011 1(12)420 360 2 11
 3.000000011011 1(12)480 432 2 16
 4.000000110011 1(12)540 504 2 21
 5.000001100011 1(12)864 792 2 24
 6.000011000011 1(6)384 360 2 25
 7.000011011011 1(12)384 360 2 20
 8.000110011011 1(12)432 408 2 24
 9.001100110011 1(12)432 408 2 24
10.001100100011 1(4)108 104 1 27
11.010101101011 1(3)48 45 1 24
  total NATs = 11
L= 13
 1.0000000000000 1(1)2 1 1 2
 2.0000000000011 1(13)546 468 2 12
 3.0000000011011 1(13)624 572 2 18
 4.0000000110011 1(13)780 728 2 24
 5.0000001100011 1(13)1170 1092 2 30
 6.0000011000011 1(13)1248 1170 2 32
 7.0000011011011 1(13)624 585 2 24
 8.0000110011011 1(13)624 598 2 30
 9.0001100110011 1(13)624 598 2 30
10.0001101101011 1(13)936 884 2 32
11.0011001101011 1(13)702 676 2 24
12.0110101101011 1(13)312 299 1 24
  total NATs = 12
L= 14
 1.00000000000000 1(1)4 2 2 2
 2.00000000000011 1(14)672 588 2 13
 3.00000000011011 1(14)840 770 2 20
 4.00000000110011 1(14)1008 952 2 27
 5.00000001100011 1(14)1680 1568 2 32
 6.00000011000011 1(14)1680 1568 2 35
 7.00000011011011 1(14)840 798 2 28
 8.00000110011011 1(7)1008 948 2 35
 9.00001100110011 1(14)1008 966 2 36
10.00011001100011 1(14)1008 966 2 36
11.00011011011011 1(14)1344 1288 2 40
12.00110011011011 1(14)1008 980 2 45
13.00110101101011 1(14)1344 1288 2 40
14.01101011011011 1(14)1512 1456 2 48
15.00110101101011 1(14)672 644 2 32
16.01101010110011 1(14)504 490 1 36
17.01101101101011 1(7)252 245 1 36
  total NATs = 17
L= 15
 1.000000000000000 1(1)2 1 1 2
 2.000000000000011 1(15)840 735 2 14
 3.000000000011011 1(15)1050 975 2 21
 4.000000000110011 1(15)1275 1200 2 30
 5.000000001100011 1(15)2160 2040 2 40
 6.000000011000011 1(15)2160 2080 2 40
 7.000000011011011 1(15)1050 1005 2 30
 8.000000110011011 1(15)2400 2280 2 40
 9.000001100110011 1(15)2700 2565 2 42
10.000011001100011 1(15)1350 1305 2 42
11.000011011011011 1(15)1350 1305 2 42
12.000110011011011 1(15)2070 2070 2 48
13.001100110011011 1(15)2160 2070 2 48
14.001101011011011 1(15)1920 1860 2 48
15.001101101101011 1(15)2160 2100 2 60
16.010101101101011 1(15)960 930 2 40
17.010110110110011 1(15)1080 1040 2 64
18.010110111011011 1(15)1080 1050 2 48
19.011011011011011 1(15)1080 1050 2 48
20.010110011011011 1(15)1080 1050 1 48
21.011010110110011 1(5)54
22.011011011011011 1(15)810 795 2 54
23.011011011011011 1(3)96 93 1 32
  total NATs = 23
```

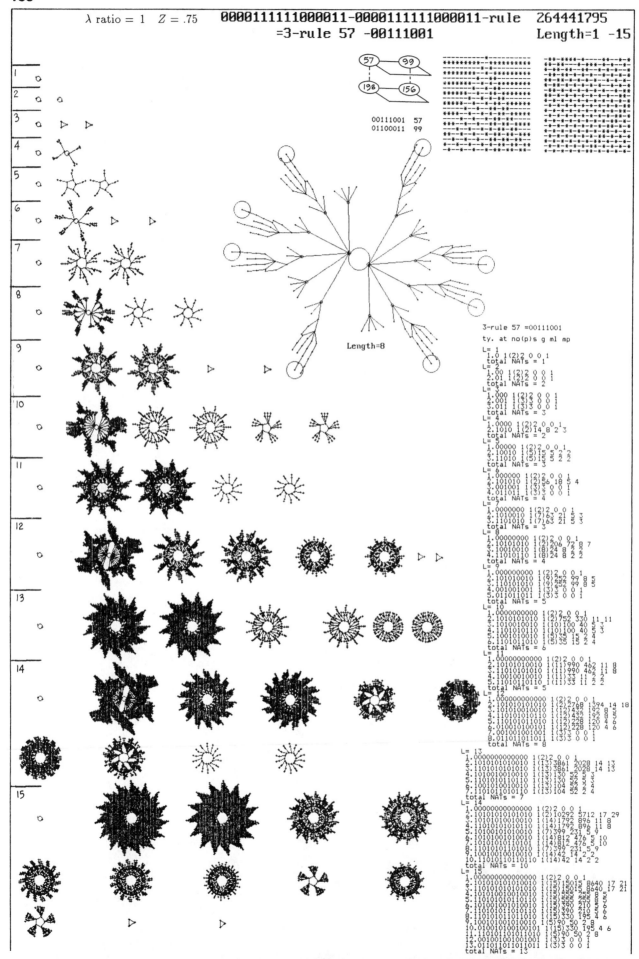

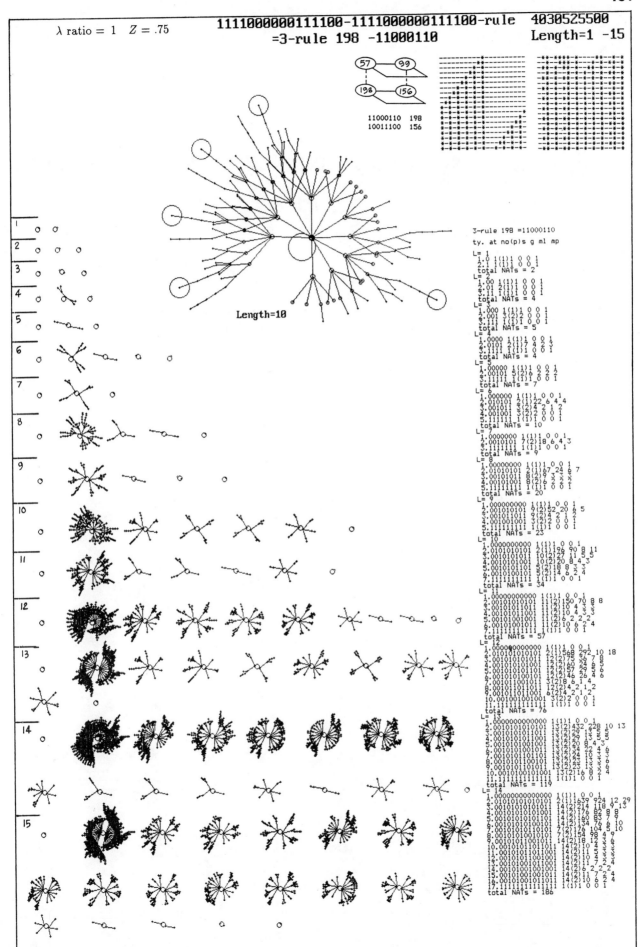

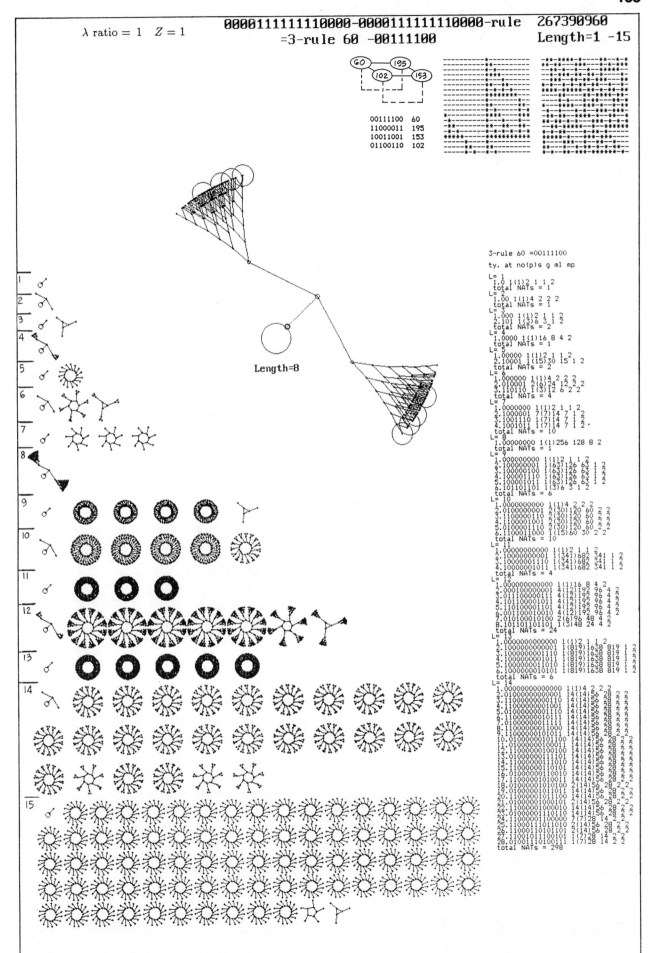

A2.4 n=5 Rules, Totalistic Code

A2.4.1 Index

Key: [decimal rule number],*[hex rule number]*-[page number]

0,*0*-85	8,*8*-193	16,*10*-189	24,*18*-197	32,*20*-187	40,*28*-195	48,*30*-191	56,*38*-199
1,*1*-186	9,*9*-200	17,*11*-212	25,*19*-220	33,*21*-223	41,*29*-219	49,*31*-211	57,*39*-197
2,*2*-188	10,*A*-202	18,*12*-215	26,*1A*-219	34,*22*-213	42,*2A*-217	50,*32*-209	58,*3A*-195
3,*3*-190	11,*B*-204	19,*13*-208	27,*1B*-200	35,*23*-210	43,*2B*-202	51,*33*-207	59,*3B*-193
4,*4*-192	12,*C*-207	20,*14*-203	28,*1C*-211	36,*24*-201	44,*2C*-209	52,*34*-205	60,*3C*-191
5,*5*-194	13,*D*-208	21,*15*-216	29,*1D*-212	37,*25*-218	45,*2D*-215	53,*35*-203	61,*3D*-189
6,*6*-196	14,*E*-210	22,*16*-218	30,*1E*-223	38,*26*-221	46,*2E*-213	54,*36*-201	62,*3E*-187
7,*7*-198	15,*F*-190	23,*17*-186	31,*1F*-186	39,*27*-196	47,*2F*-188	55,*37*-192	63,*3F*-85

A2.4.2 Totalistic Code Clusters (see section 3.3.10)

$C = C_r$, so the reflection links collapse.

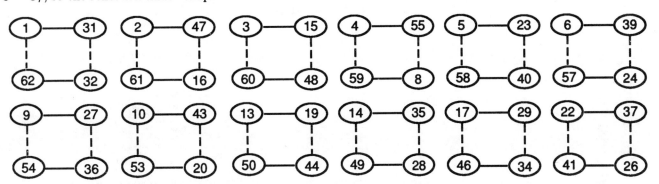

The clusters collapse further where for a given rule C, $C_c = C_n$

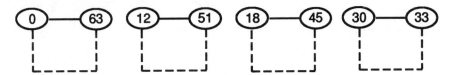

and also if $C = C_n$

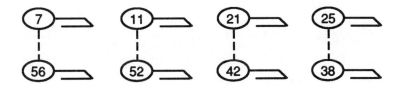

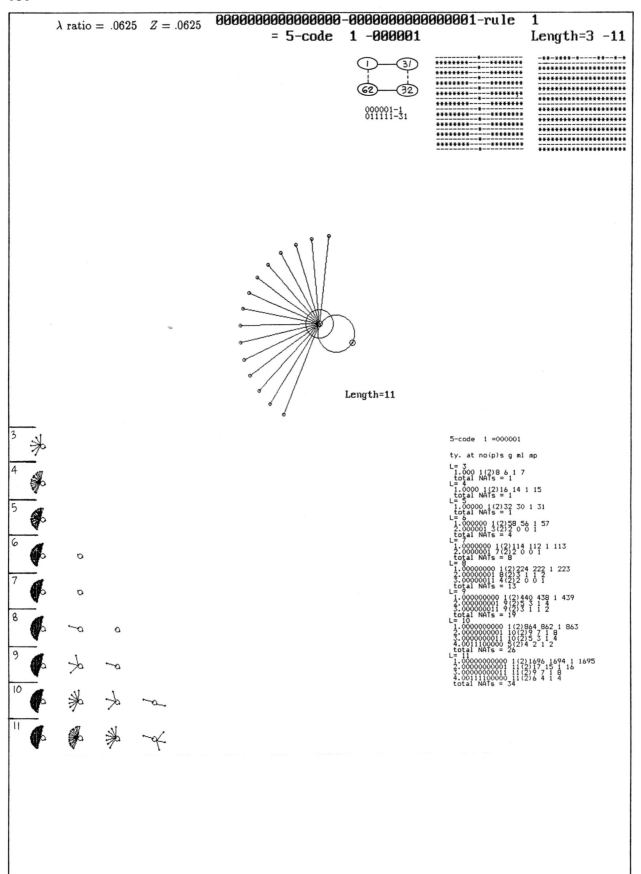

187

λ ratio = .0625   Z = .0625   1111111111111111-1111111111111110-rule   4294967294
                              = 5-code 62 -111110   Length=3 -11

111110-62
100000-32

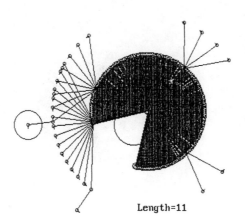

Length=11

3
4
5
6
7
8
9
10
11

```
5-code  62 =111110
ty. at no(p)s g ml mp
L= 3
 1.000    1(1)1   0 0 1
 2.111    1(1)7   6 1 7
 total NATs = 2
L= 4
 1.0000   1(1)1   0 0 1
 2.1111   1(1)15 14 1 15
 total NATs = 2
L= 5
 1.00000  1(1)1   0 0 1
 2.11111  1(1)31 30 1 31
 total NATs = 2
L= 6
 1.000000 1(1)1   0 0 1
 2.111111 1(1)63 56 2 57
 total NATs = 2
L= 7
 1.0000000 1(1)1   0 0 1
 2.1111111 1(1)127 112 2 113
 total NATs = 2
L= 8
 1.00000000 1(1)1   0 0 1
 2.11111111 1(1)255 230 2 223
 total NATs = 2
L= 9
 1.000000000 1(1)1   0 0 1
 2.111111111 1(1)511 474 2 439
 total NATs = 2
L= 10
 1.0000000000 1(1)1   0 0 1
 2.1111111111 1(1)1023 972 3 863
 total NATs = 2
L= 11
 1.00000000000 1(1)1   0 0 1
 2.11111111111 1(1)2047 1980 3 1695
 total NATs = 2
L= 12
```

188

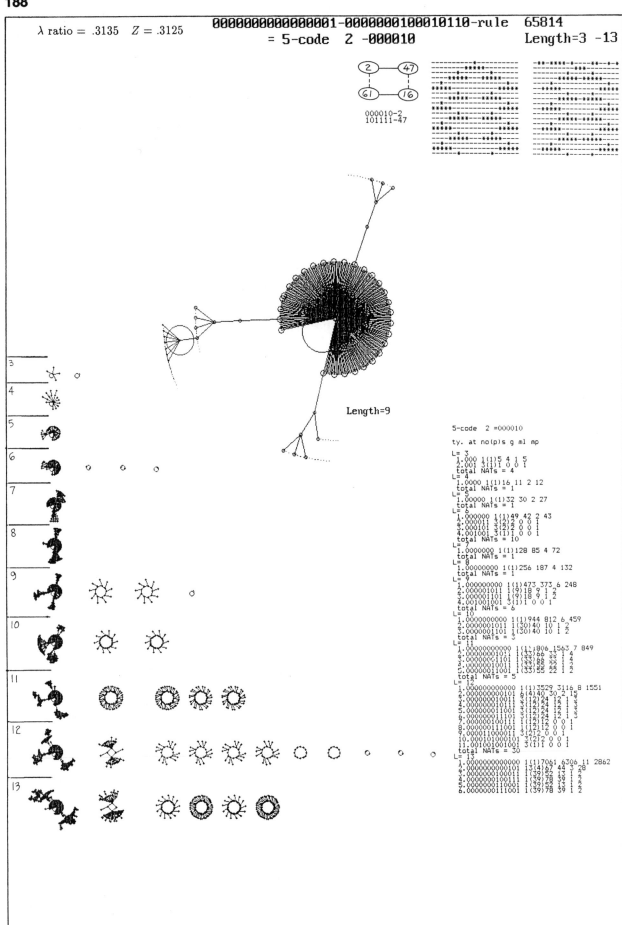

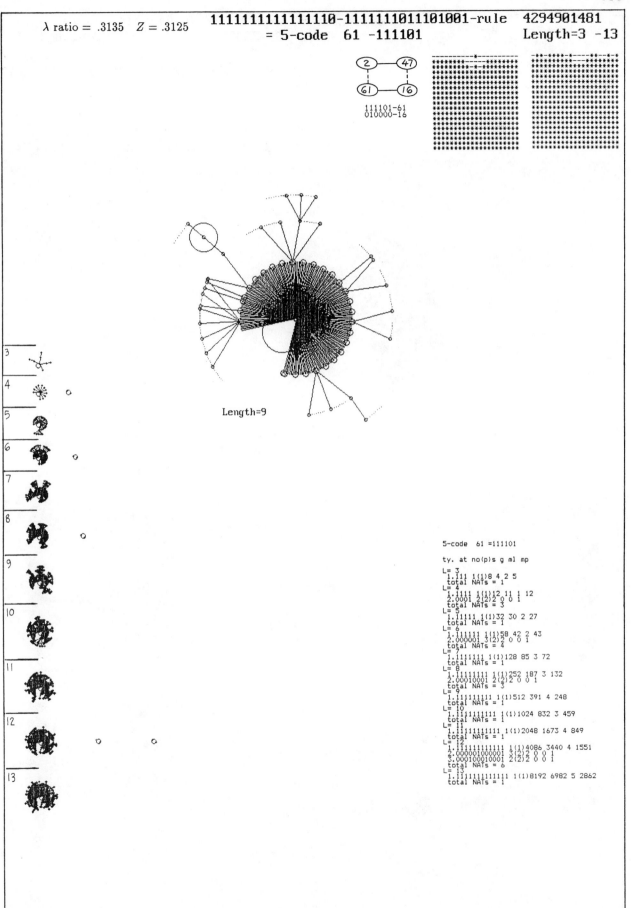

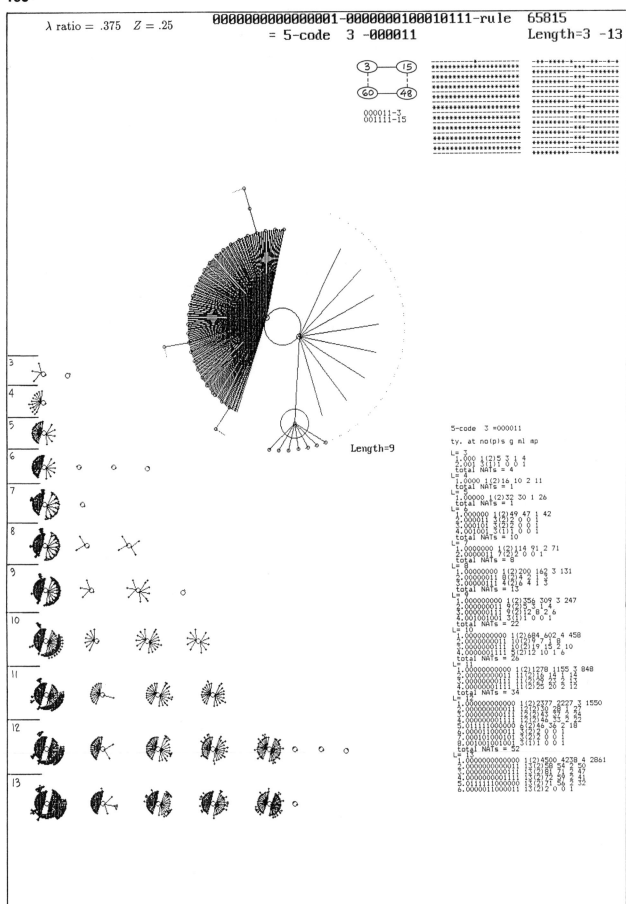

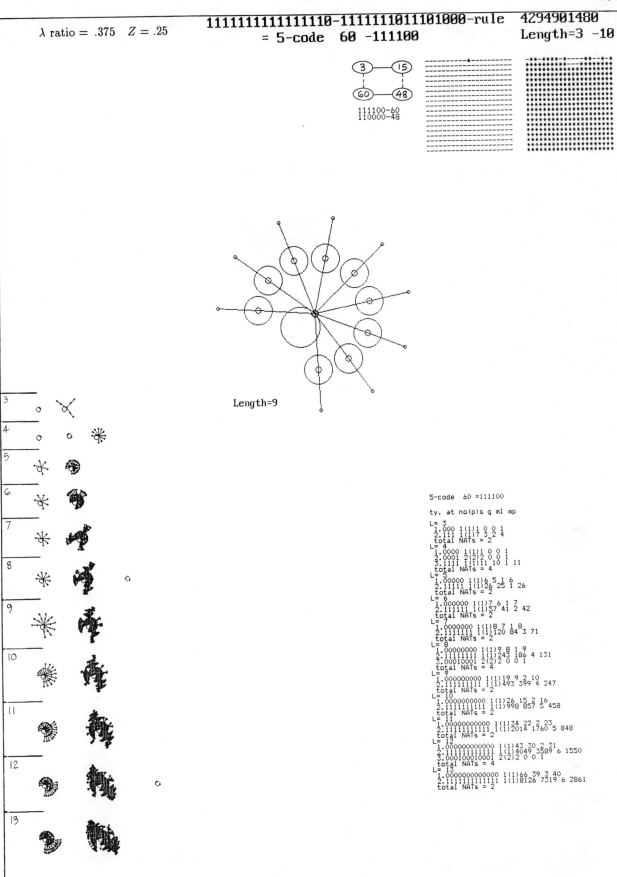

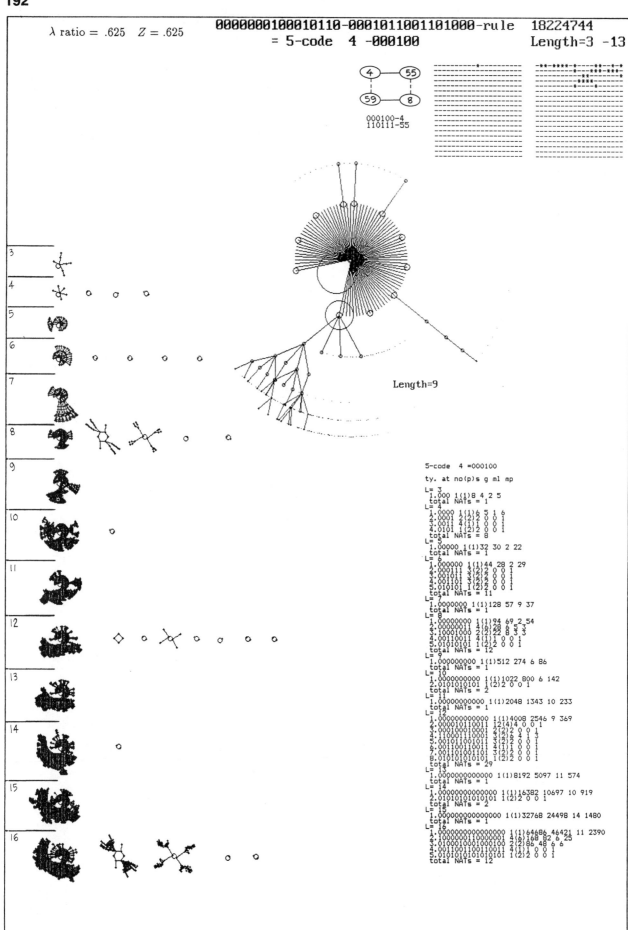

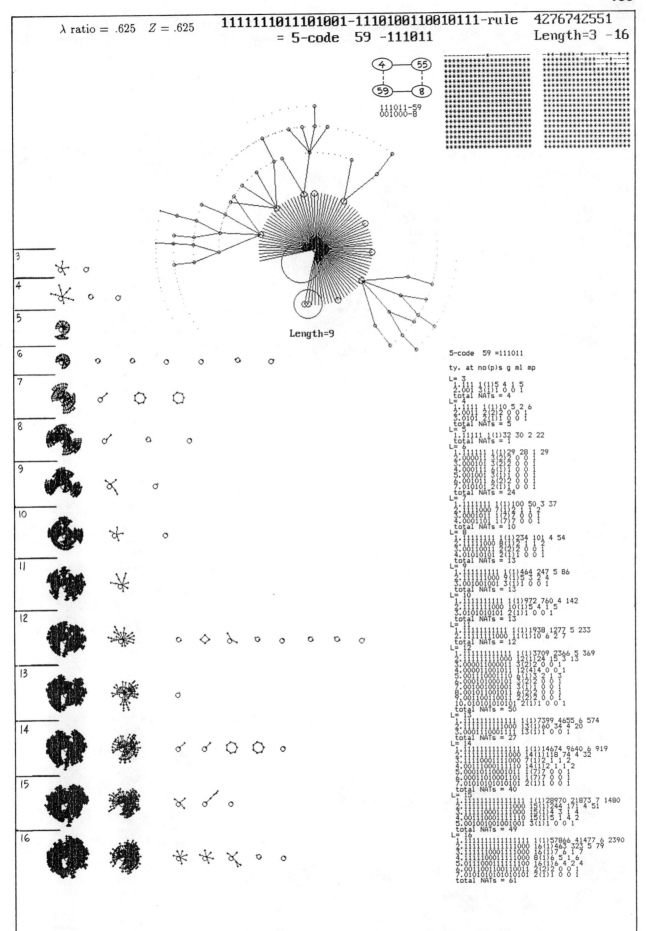

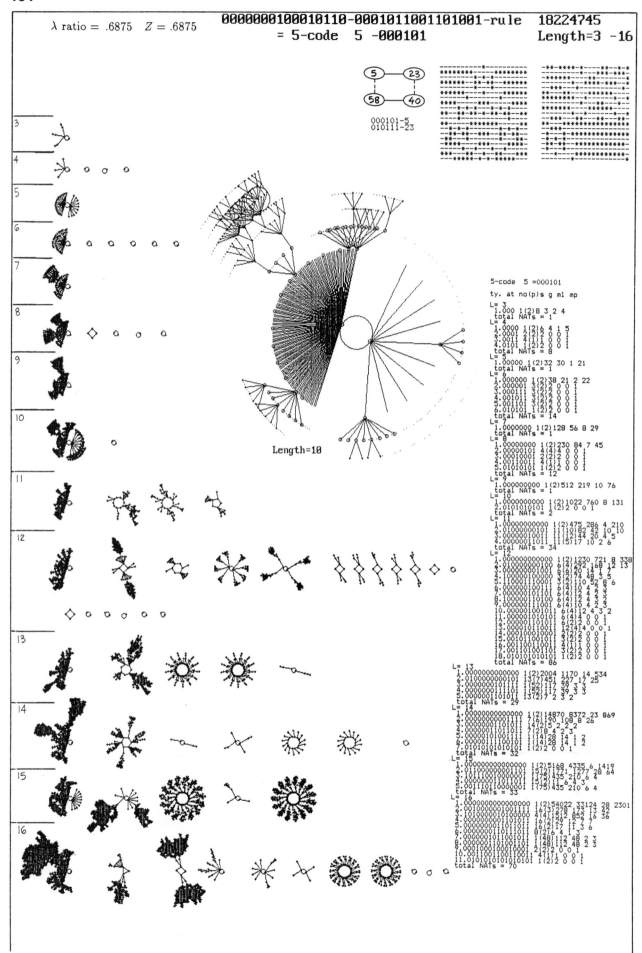

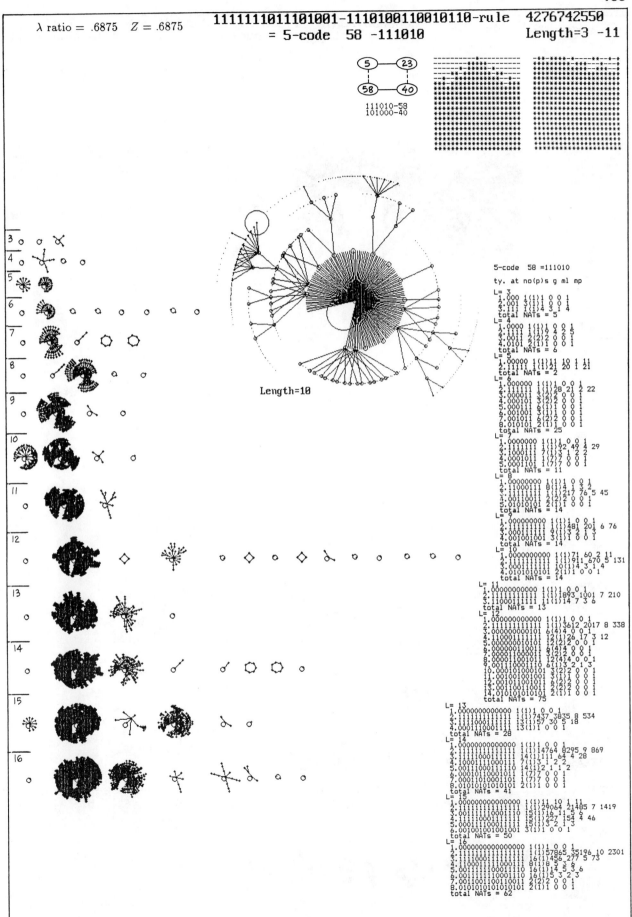

196

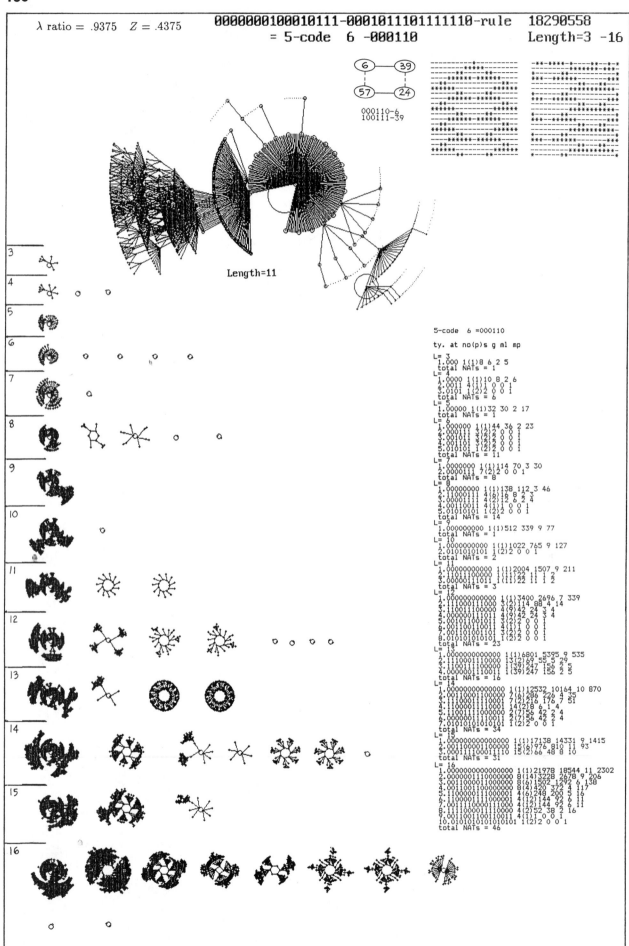

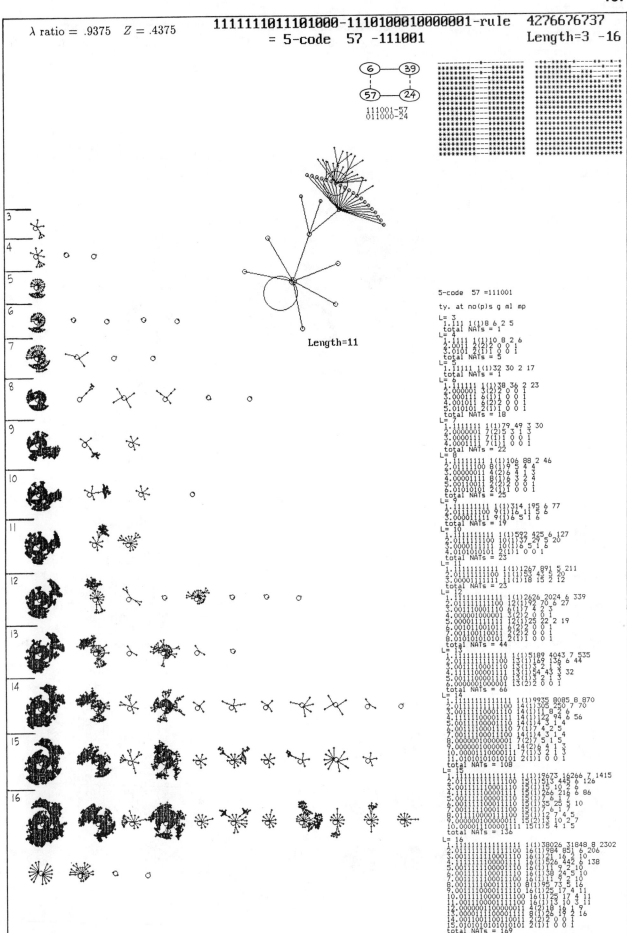

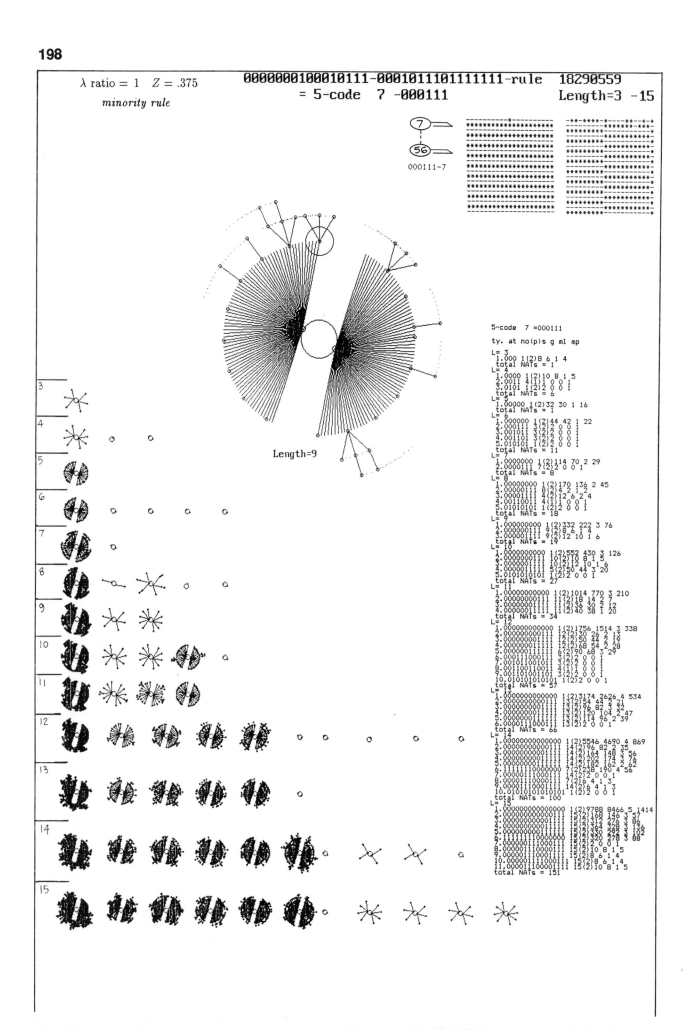

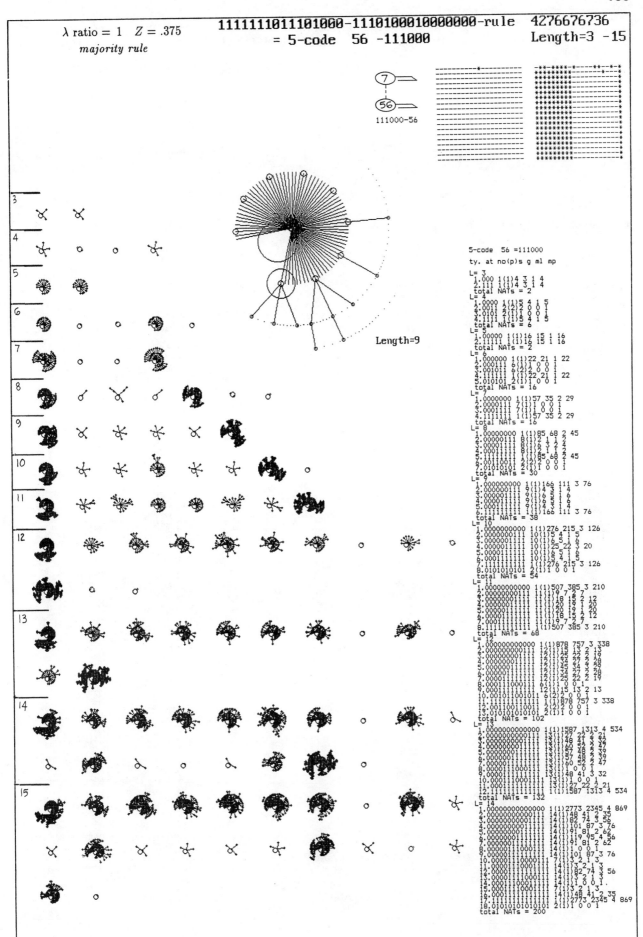

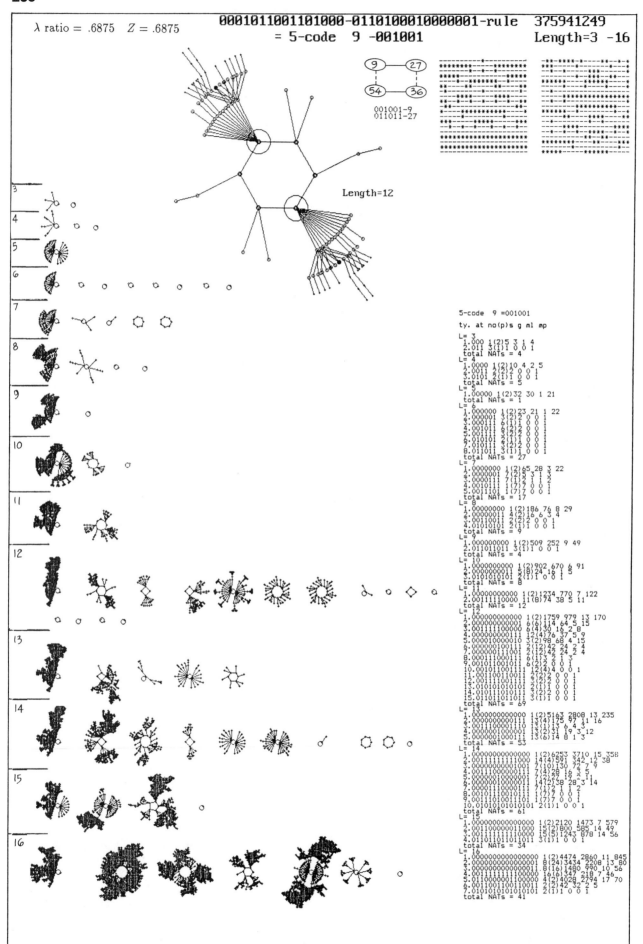

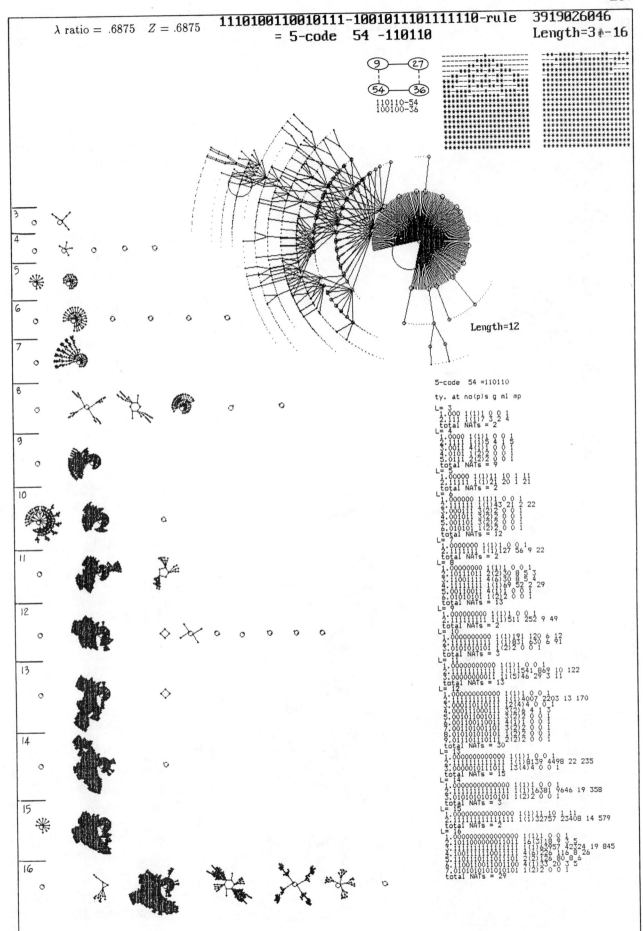

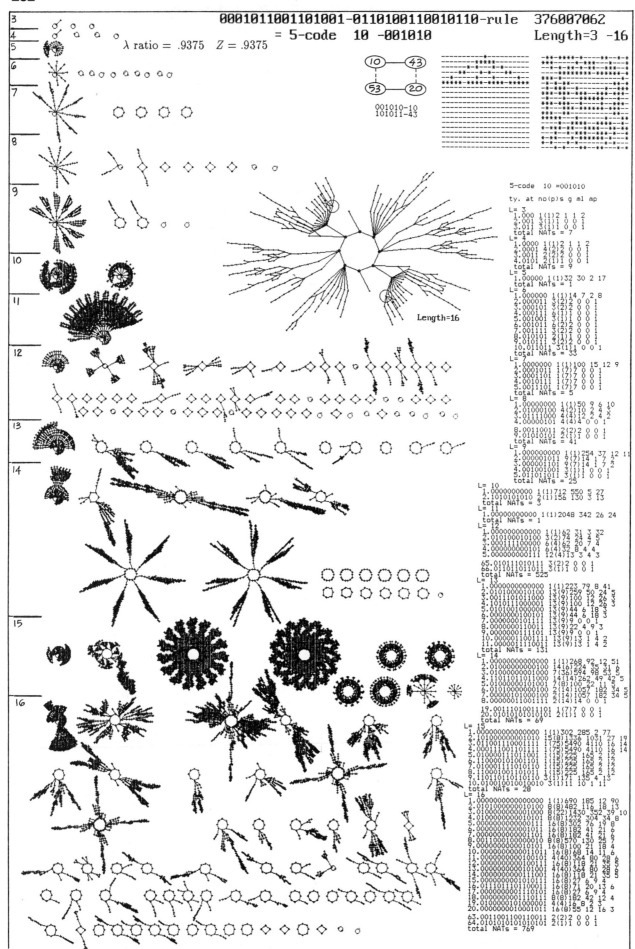

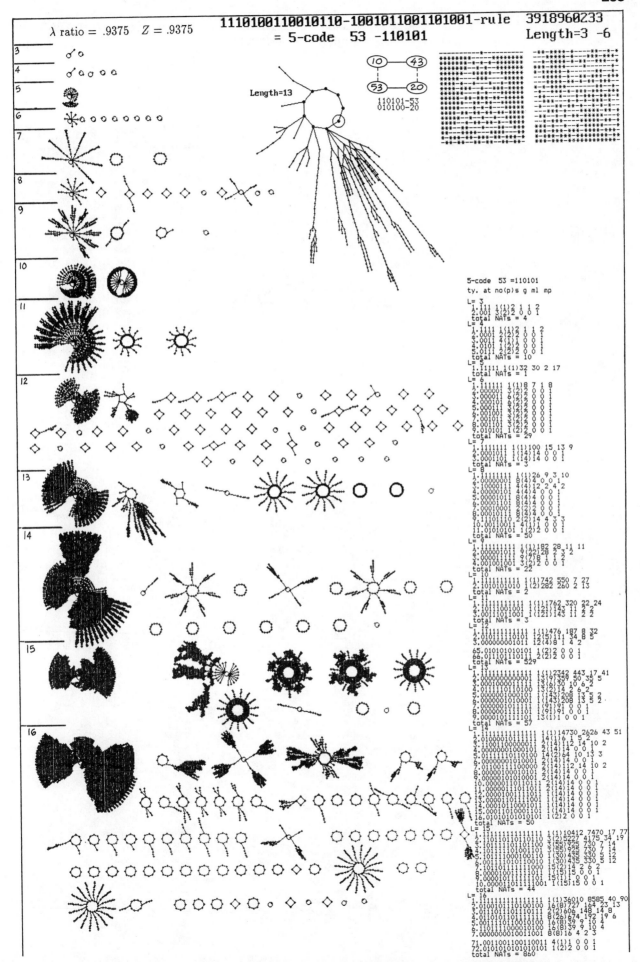

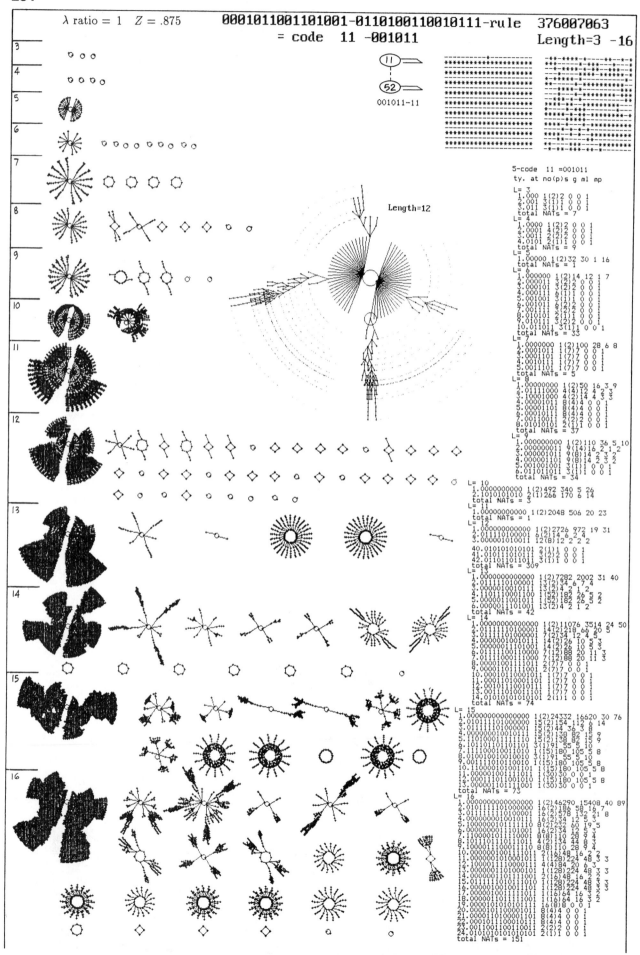

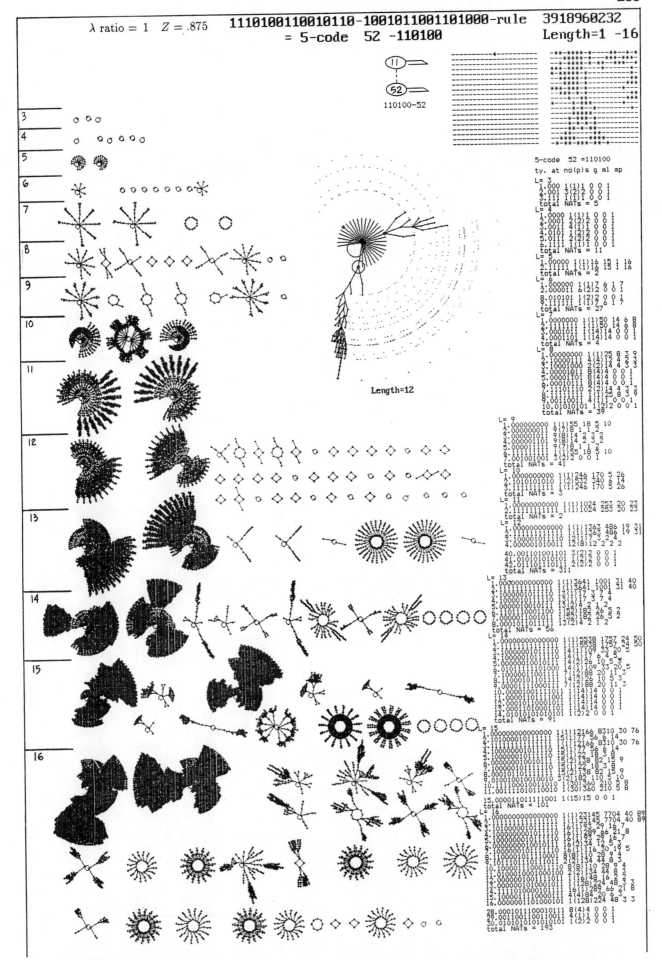

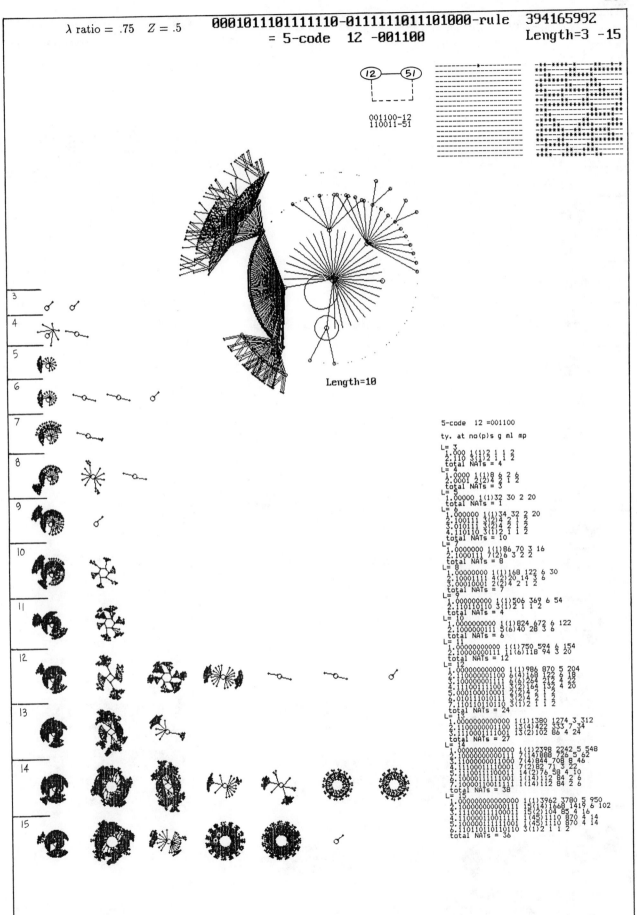

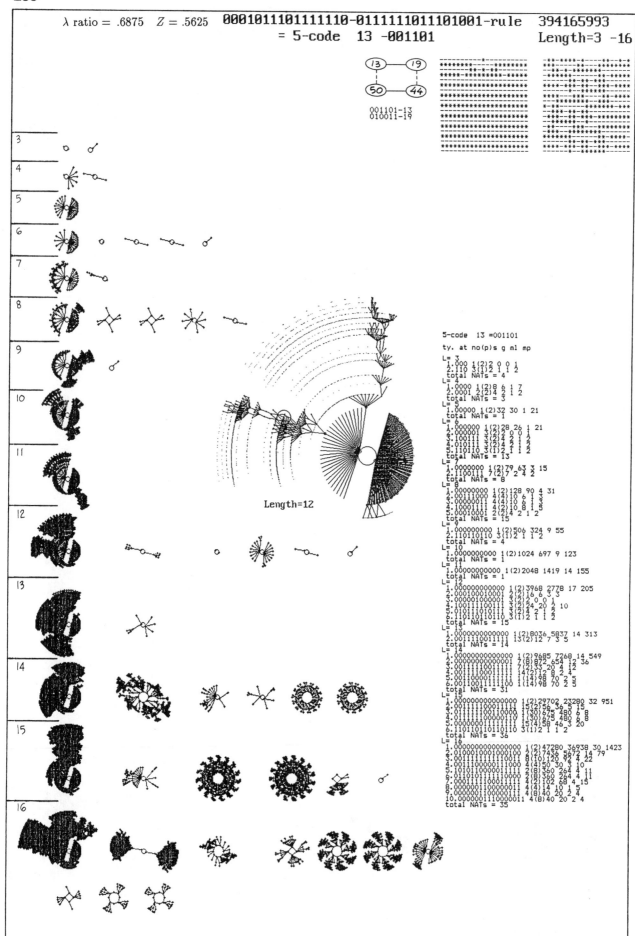

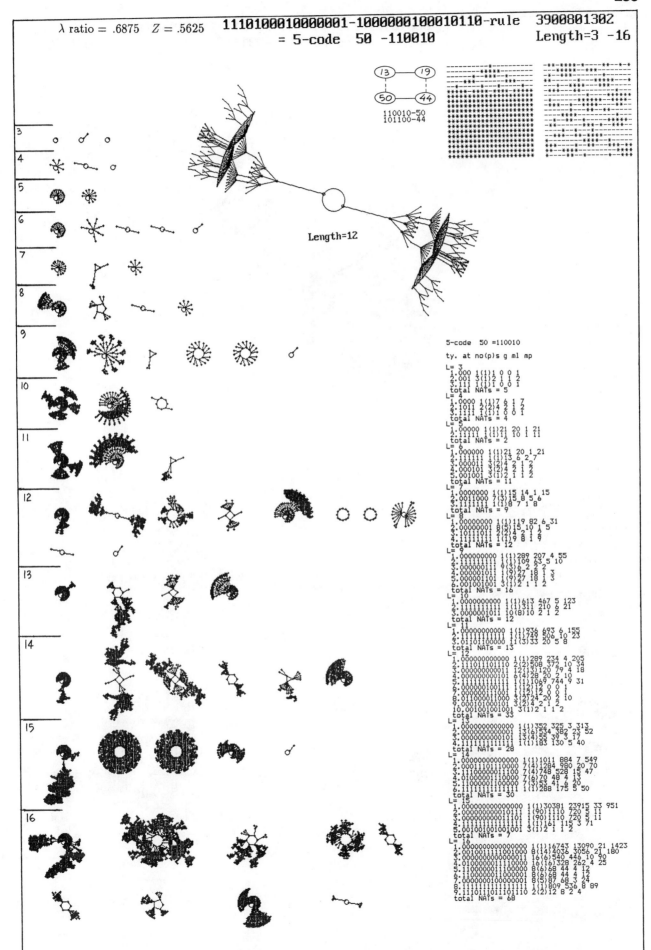

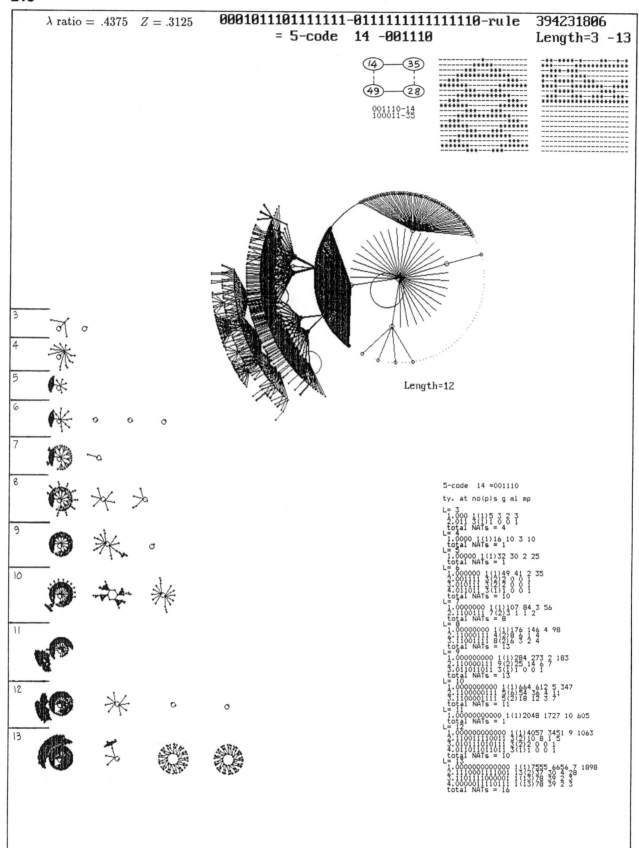

λ ratio = .4375   Z = .3125   **1110100010000000-1000000000000001**-rule   3900735489
= 5-code   49 -110001   Length=3 -13

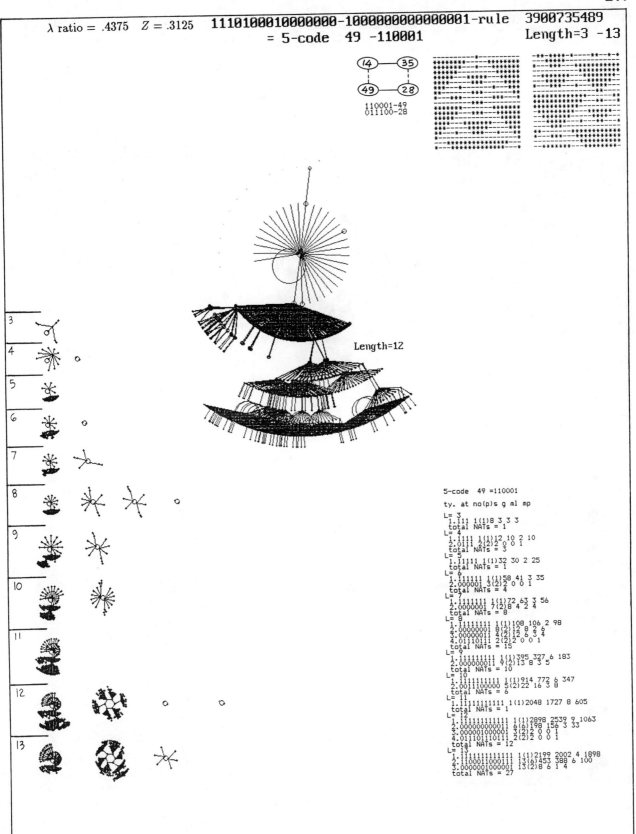

Length=12

```
5-code  49 =110001
ty. at no(p)s g ml mp
L= 3
 1.111 1(1)8 3 3 3
 total NATs = 1
L= 4
 1.1111 1(1)12 10 2 10
 2.0111 2(2)2 0 0 1
 total NATs = 3
L= 5
 1.11111 1(1)32 30 2 25
 total NATs = 1
L= 6
 1.111111 1(1)58 41 3 35
 2.000001 3(2)2 0 0 1
 total NATs = 4
L= 7
 1.1111111 1(1)72 63 3 56
 2.0000001 7(2)8 4 2 4
 total NATs = 8
L= 8
 1.11111111 1(1)108 106 2 98
 2.00000001 8(2)12 8 2 6
 3.00000011 4(2)12 6 3 4
 4.01110111 2(2)2 0 0 1
 total NATs = 15
L= 9
 1.111111111 1(1)395 327 6 183
 2.000000011 9(2)13 8 3 5
 total NATs = 10
L= 10
 1.1111111111 1(1)914 772 6 347
 2.0011100000 5(2)22 16 3 8
 total NATs = 6
L= 11
 1.11111111111 1(1)2048 1727 8 605
 total NATs = 1
L= 12
 1.111111111111 1(1)2898 2539 9 1063
 2.000000000011 6(6)198 156 3 33
 3.000001000001 3(2)2 0 0 1
 4.011101110111 2(2)2 0 0 1
 total NATs = 12
L= 13
 1.1111111111111 1(1)2199 2002 4 1898
 2.1100010000111 13(6)453 388 6 100
 3.0000001000001 13(2)8 6 1 4
 total NATs = 27
```

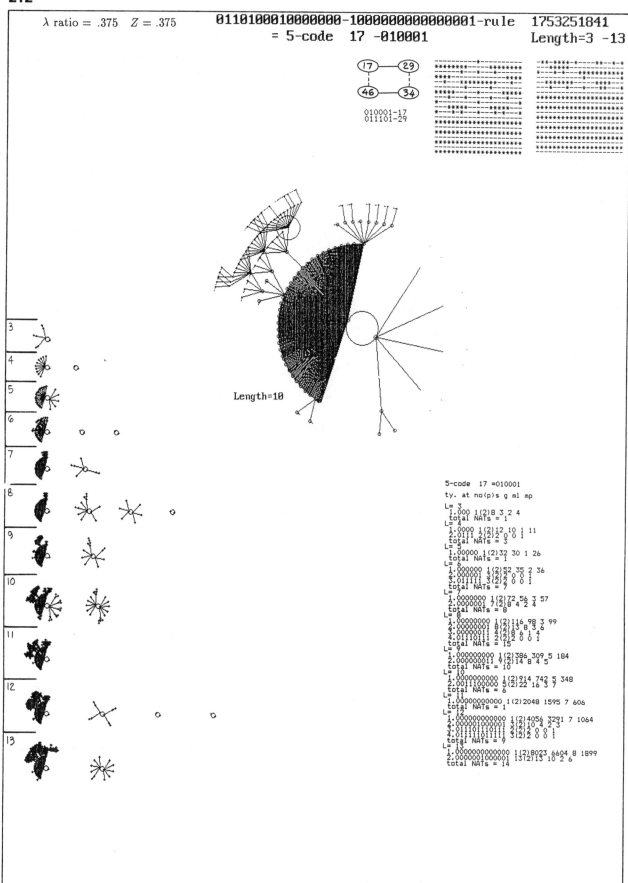

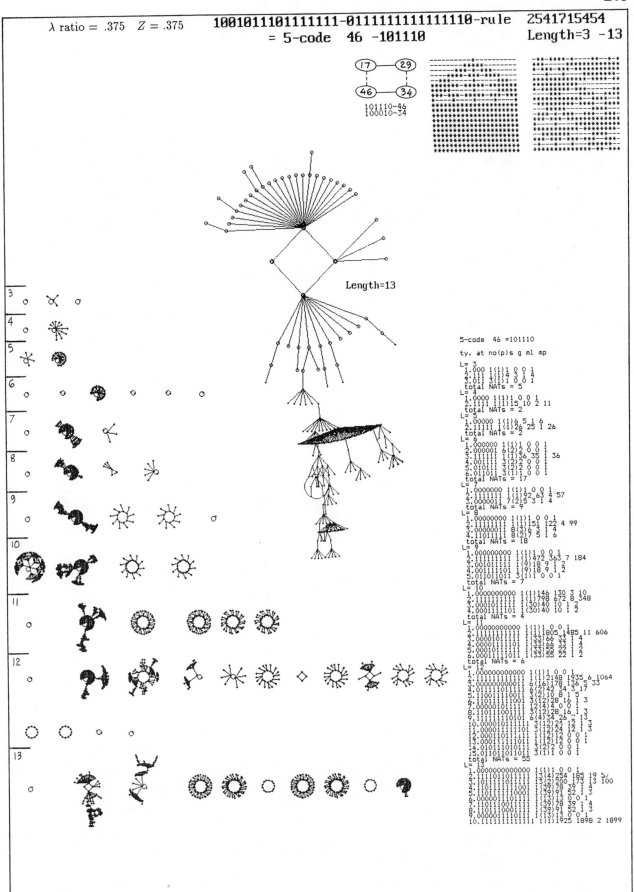

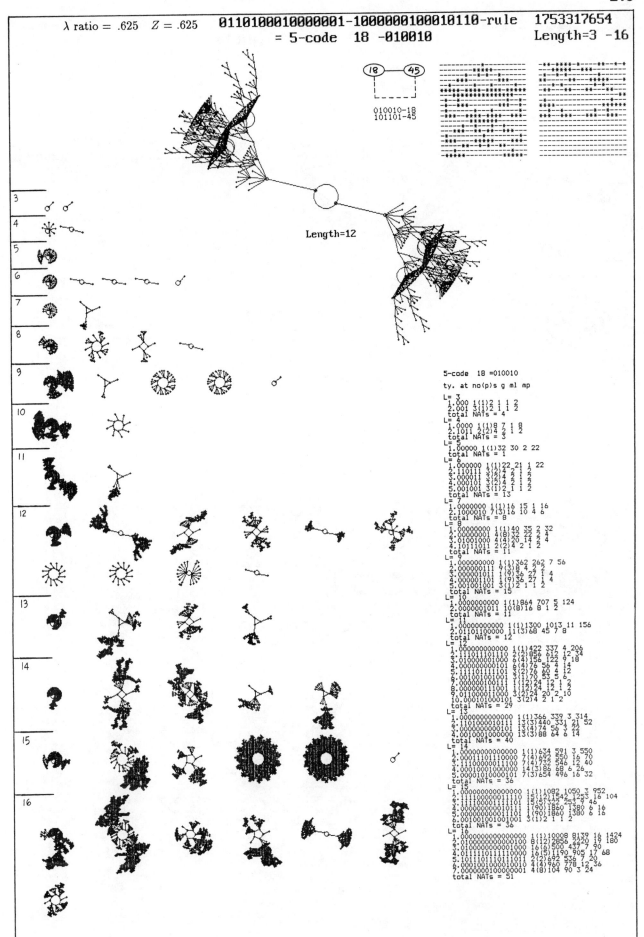

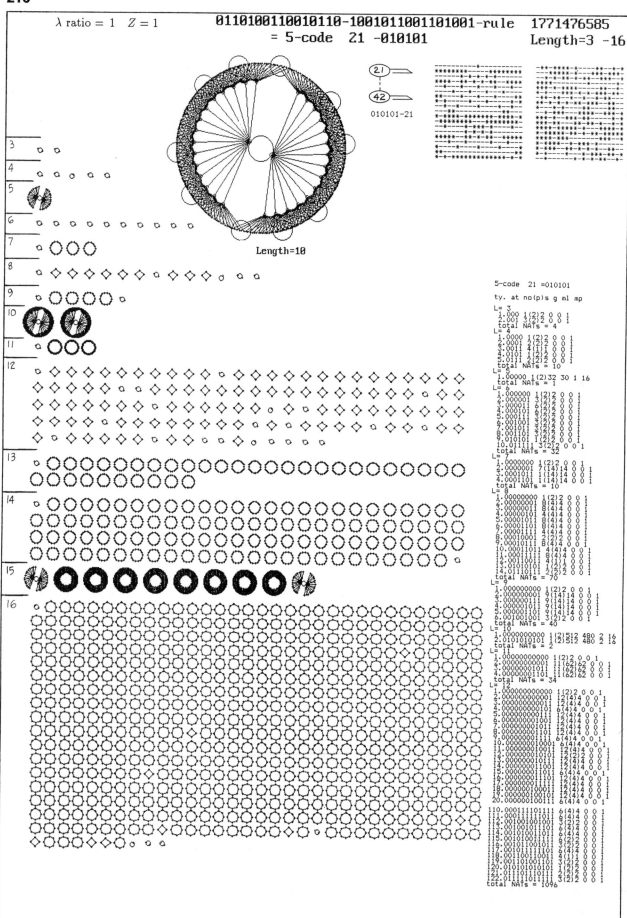

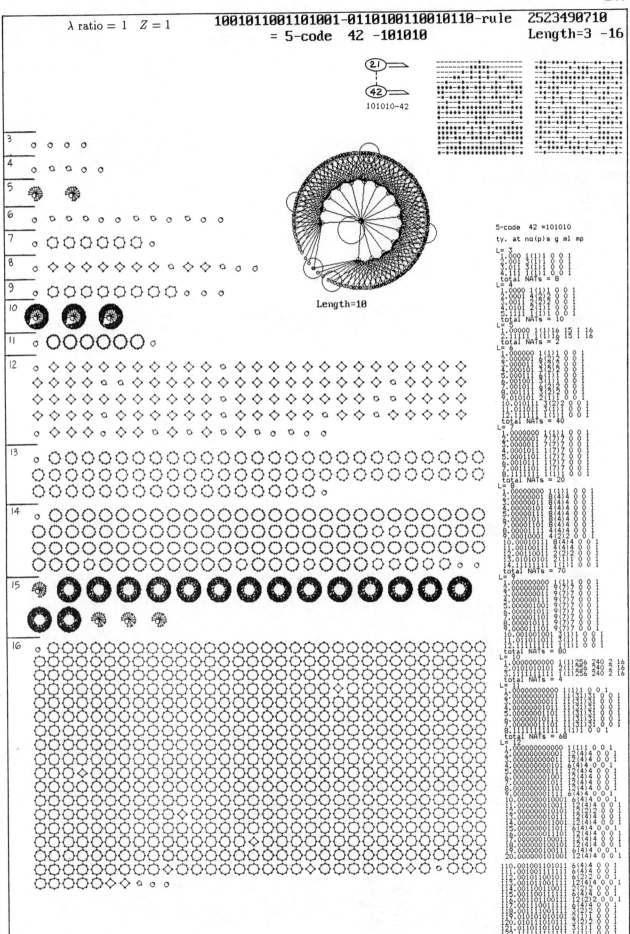

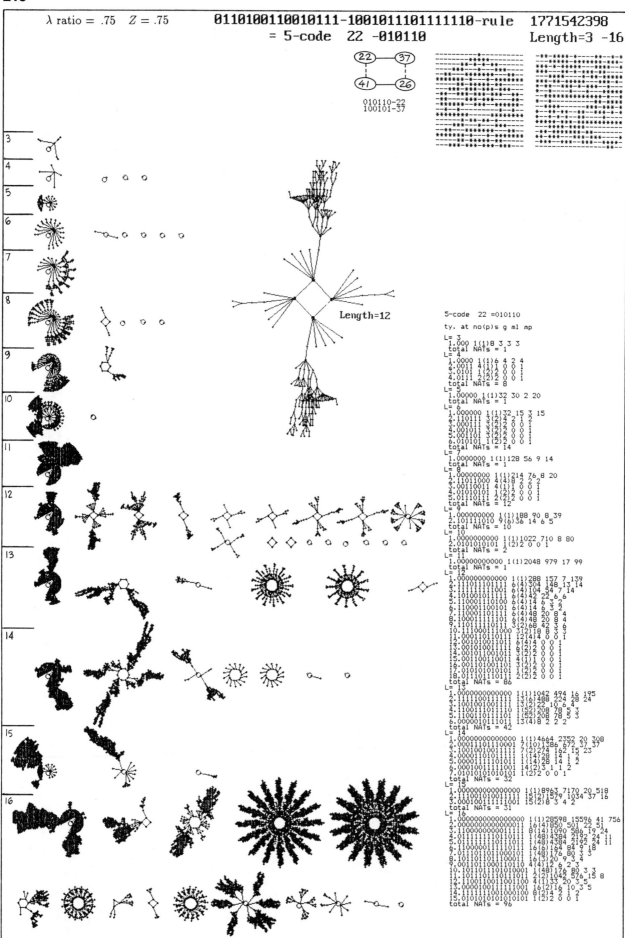

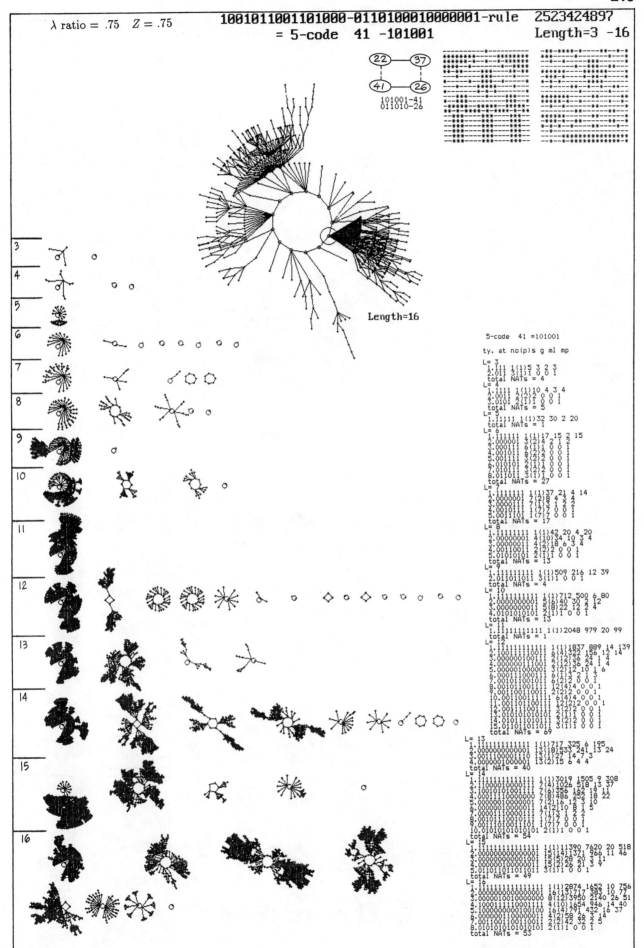

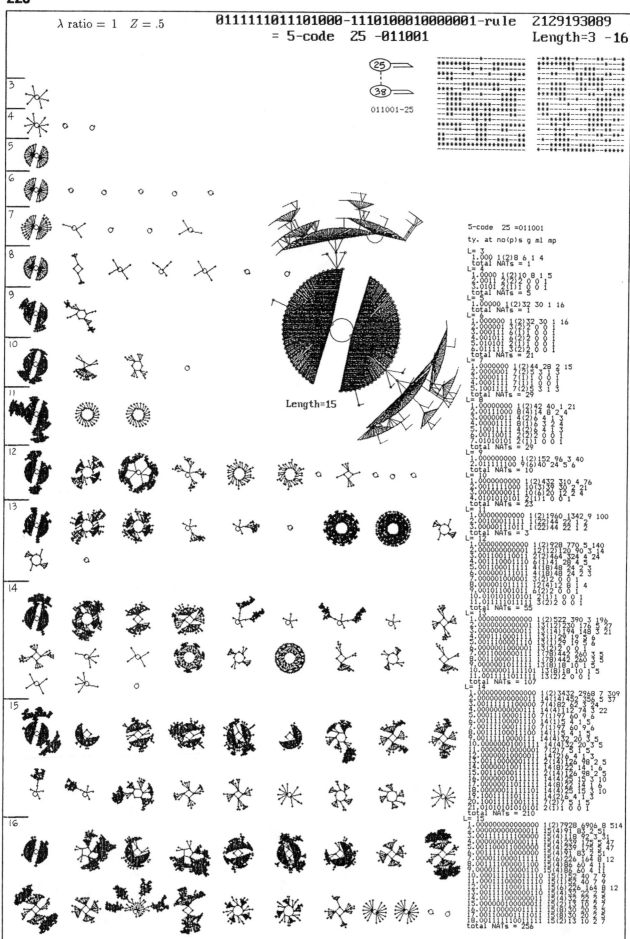

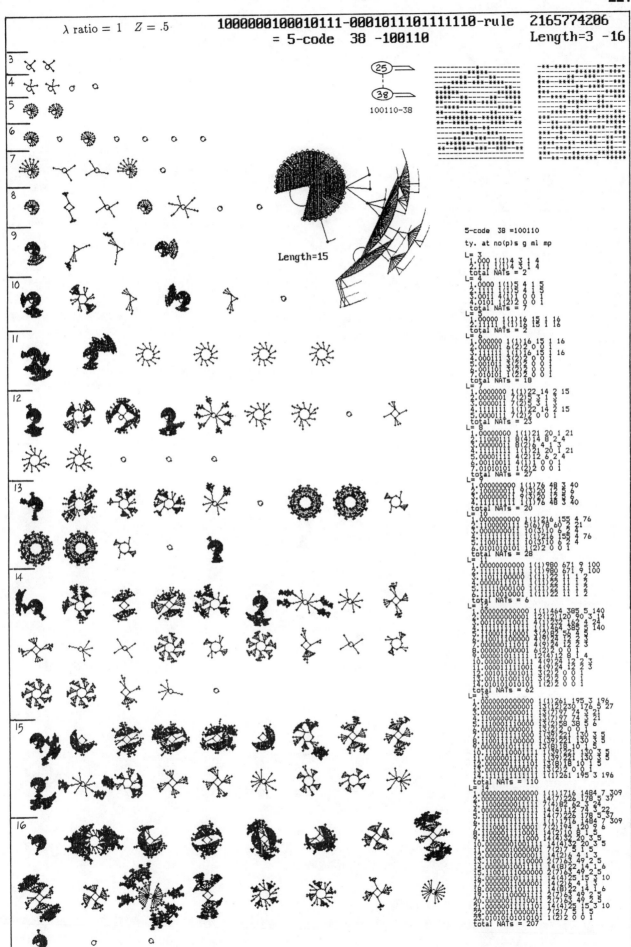

λ ratio = .125 Z = .125 0111111111111111-1111111111111110-rule 2147483646
= 5-code 30 -011110 Length=3 -12

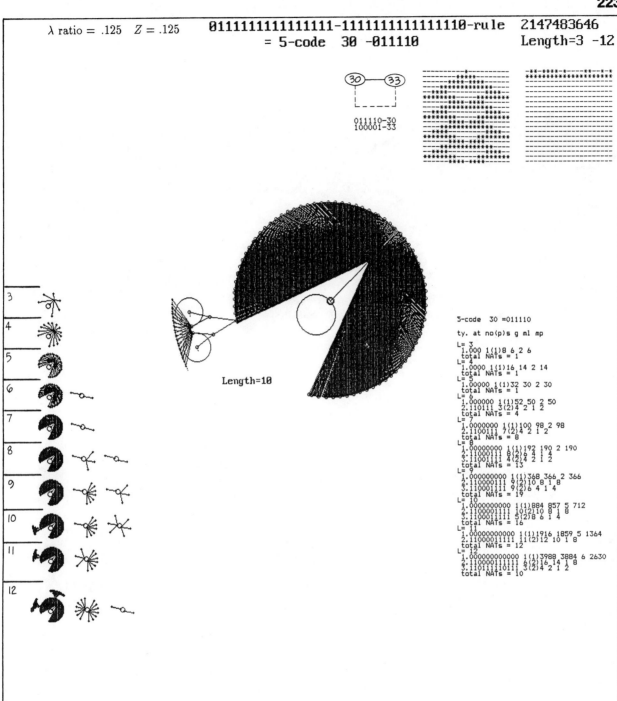

APPENDIX 3
Mutants

A3.1 The Effect of Mutating a Rule on the Basin of Attraction (see section 4.2)

Diagrams A3.1-16 show a sequence of basins of attraction (or basin fields) corresponding to a sequence of mutated rules.

The mutations are made to the rule's $n = 5$ rule table, even if the rule is an $n = 3$ rule ($n = 3$ rules can be expressed as $n = 5$ rules; see 3.3.10). The rule table consists of a 32-bit string arranged in conventional order (see 3.3.9). The mutations are made to successive bits in the rule table from left to right. The series of mutated rules consists of a source rule, and 32 one-bit mutants. In diagrams A3.2 and A3.10 only, the one-bit mutations are cumulative, so that the source rule is progressively changed to its complement.

In the diagrams, the source rule is located in the botton left-hand corner frame (bold outline); successive mutants are located to the right of the source rule, and then from left to right in ascending rows.

Diagrams A3.1-2 show the basin of attraction field. Diagrams A3.3-16 show single basins.

The rule table may be thought of as analogous to a genetic code or *genotype*, and the basin structure as the resulting *phenotype*. Small mutations to the genotype result, in general, to small changes in the phenotype.

APPENDIX 3 Mutants

A3.1

rule 60, $L = 8$, NAT field
one-bit mutants

A3.2

rule 60, $L = 8$, NAT field
one-bit cumulative mutants

A3.1 The Effect of Mutating a Rule on the Basin of Attraction

A3.3

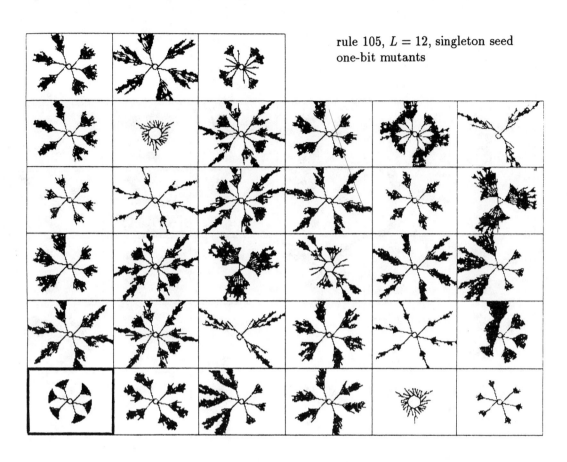

rule 105, $L = 12$, singleton seed
one-bit mutants

A3.4

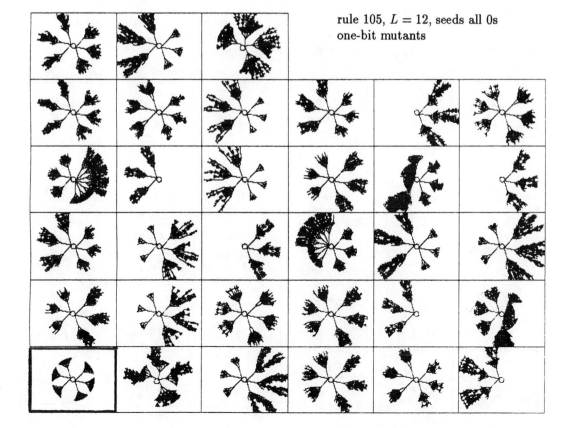

rule 105, $L = 12$, seeds all 0s
one-bit mutants

228   APPENDIX 3 Mutants

A3.5

rule 105, $L = 12$, seeds all 0s
one-bit mutants

A3.6

rule 9, $L = 8$, singleton seed
one-bit mutants

A3.1 The Effect of Mutating a Rule on the Basin of Attraction

A3.7

rule 228, $L = 11$, singleton seed
one-bit mutants

A3.8

rule 9, $L = 10$, singleton seed
one-bit mutants

230 APPENDIX 3 Mutants

A3.9

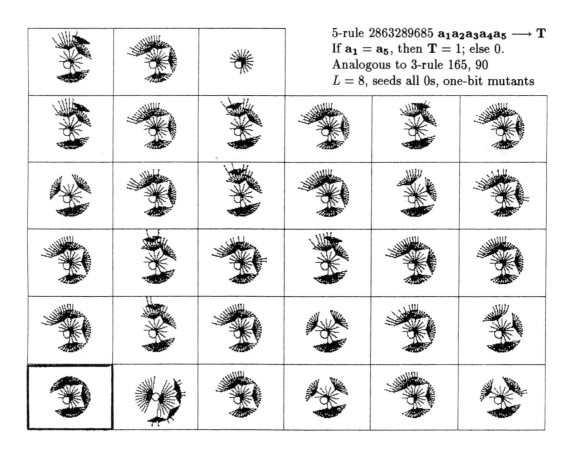

5-rule 2863289685 $a_1 a_2 a_3 a_4 a_5 \longrightarrow T$
If $a_1 = a_5$, then $T = 1$; else 0.
Analogous to 3-rule 165, 90
$L = 8$, seeds all 0s, one-bit mutants

A3.10

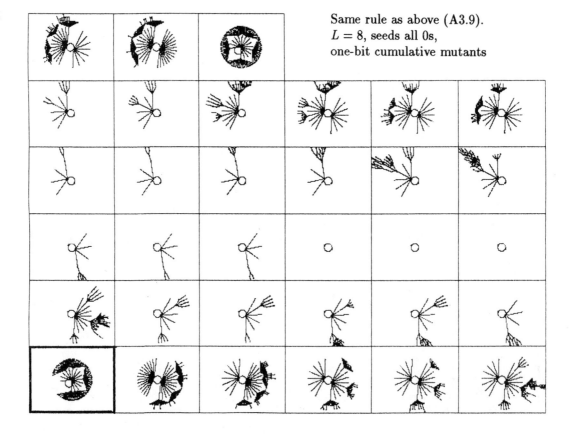

Same rule as above (A3.9).
$L = 8$, seeds all 0s,
one-bit cumulative mutants

A3.1 The Effect of Mutating a Rule on the Basin of Attraction

A3.11

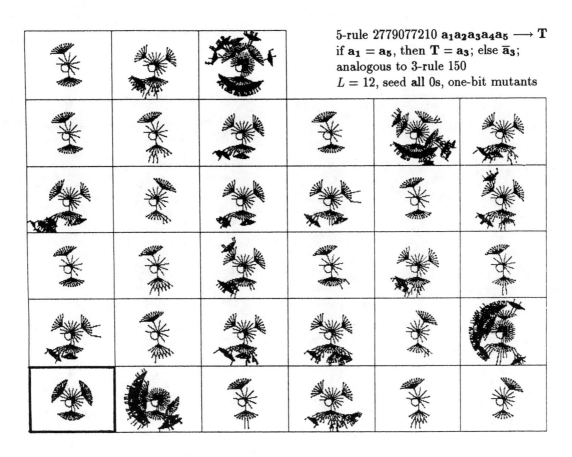

5-rule 2779077210 $a_1a_2a_3a_4a_5 \longrightarrow T$
if $a_1 = a_5$, then $T = a_3$; else $\bar{a}_3$;
analogous to 3-rule 150
$L = 12$, seed all 0s, one-bit mutants

A3.12

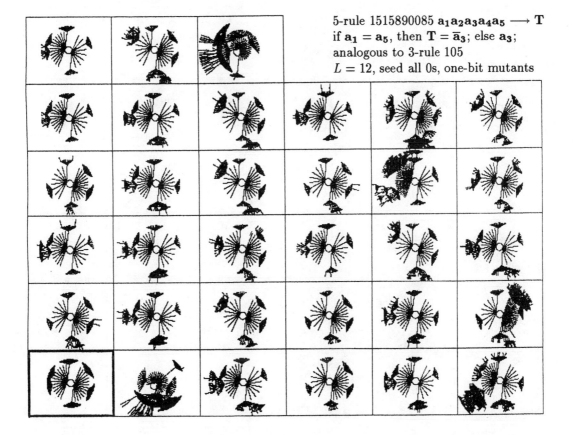

5-rule 1515890085 $a_1a_2a_3a_4a_5 \longrightarrow T$
if $a_1 = a_5$, then $T = \bar{a}_3$; else a_3;
analogous to 3-rule 105
$L = 12$, seed all 0s, one-bit mutants

APPENDIX 3 Mutants

A3.13

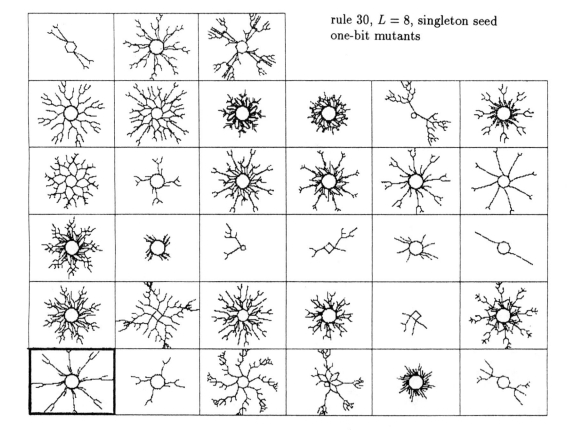

rule 25, $L = 9$, singleton seed
one-bit mutants

A3.14

rule 30, $L = 8$, singleton seed
one-bit mutants

A3.1 The Effect of Mutating a Rule on the Basin of Attraction

A3.15

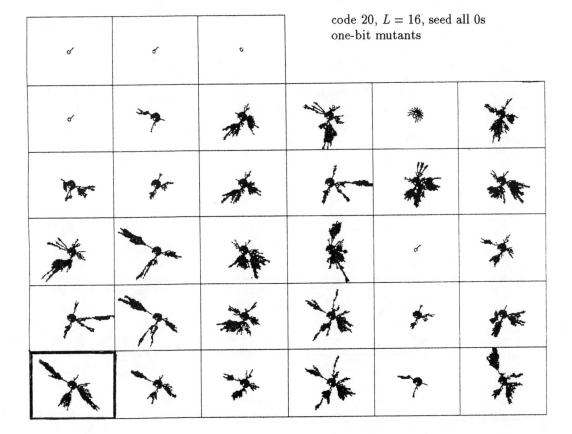

rule 193, $L = 5$, singleton seed
one-bit mutants

A3.16

code 20, $L = 16$, seed all 0s
one-bit mutants

APPENDIX 4
The Rule-Space Matrix, n=3 Rules

The 256 $n = 3$ rules (0 to 255) may be set out on a 16×16 matrix. Rows i and columns j are numbered 0 to 15 as shown on the next page, and equivalently by the 4-bit binary numbers 0000 to 1111. Each entry in the matrix, a_{ij}, is a function of its position, and is assigned the decimal equivalent of the 8-bit binary number formed by the concatenation (denoted by the symbol +) of its 4-bit row and column binary expressions. If bin\$ (x) = the 4-bit binary string equivalent of x, and dec(x\$) = the decimal equivalent of the binary string x\$, then

$$a_{ij} = \text{dec}(\text{bin\$}(i) + \text{bin\$}(j));$$

for example, to establish the entry $a_{5,6}$

bin\$(5) = "0101", and bin\$(6) = "0110"
bin\$(5) + bin\$(6) = "0101" + "0110" = "01010110", dec("01010110") = 86

Conversely, given a rule number, 0 to 255, its position on the matrix is found by the separating its 8-bit binary expression into two equal parts. The left 4 bits denotes the row i and the right 4 bits the column j. The resulting matrix is set out on the next page.

APPENDIX 4 The Rule-Space Matrix, n=3 Rules

	j															
	0	1	2	3	4	5	6	7	8	9	10	11	12	13	14	15
0	0	1	2	3	4	5	6	7	8	9	10	11	12	13	14	15
1	16	17	18	19	20	21	22	23	24	25	26	27	28	29	30	21
2	32	33	34	35	36	37	38	39	40	41	42	43	44	45	46	47
3	48	49	50	51	52	53	54	55	56	57	58	59	60	61	62	63
4	64	65	66	67	68	69	70	71	72	73	74	75	76	77	78	79
5	80	81	82	83	84	85	86	87	88	89	90	91	92	93	94	95
6	96	97	98	99	100	101	102	103	104	105	106	107	108	109	110	111
7	112	113	114	115	116	117	118	119	120	121	122	123	124	125	126	127
8	128	129	130	131	132	133	134	135	136	137	138	139	140	141	142	143
9	144	145	146	147	148	149	150	151	152	153	154	155	156	157	158	159
10	160	161	162	163	164	165	166	167	168	169	170	171	172	173	174	175
11	176	177	178	179	180	181	182	183	184	185	186	187	188	189	190	191
12	192	193	194	195	196	197	198	199	200	201	202	203	204	205	206	207
13	208	209	210	211	212	213	214	215	216	217	218	219	220	221	222	223
14	224	225	226	227	228	229	230	231	232	233	234	235	236	237	238	239
15	240	241	242	243	244	245	246	247	248	249	250	251	252	253	254	255

The position of rules on the matrix depends on the rule numbering convention implicit in the sequence of the rule table entries. The conventional sequence as described in section 3.3 is as follows:

$$\begin{array}{ccccccccc} & 111 & 110 & 101 & 100 & 011 & 010 & 001 & 000 & \text{neighbourhoods} \\ \text{Rule table..} & T_7 & T_6 & T_5 & T_4 & T_3 & T_2 & T_1 & T_0 & \text{outputs} \end{array}$$

Thus $T_7T_6T_5T_4T_3T_2T_1T_0$ is the conventional binary expression of the rule. Other sequences of the rule table would be equally valid; indeed, there are 8! = 13440 permutations, and thus the same number of possible alternative numbering conventions.

If the equivalence relationships are indicated by drawing lines between equivalent rules on the matrix, a systematic pattern is apparent, however the clarity of this pattern varies for different numbering conventions. A limited search of alternatives has turned up a permutation that results in a pattern of exceptional clarity; this is the rearranged sequence:

$$\begin{array}{ccccccccc} & 111 & 101 & 110 & 100 & 000 & 010 & 001 & 011 & \text{neighbourhoods} \\ \text{Rule table..} & T_7 & T_5 & T_6 & T_4 & T_0 & T_2 & T_1 & T_3 & \text{outputs} \end{array}$$

giving the alternative binary expression of the rule $T_7T_5T_6T_4T_0T_2T_1T_3$. In the matrix shown on the next page, rules are *positioned* according to the alternative numbering system, but are still numbered according to the conventional system. The diagonals, labeled c (complement) and n (negative), are indicated.

The matrix shows reflection equivalent rules linked by lines, and provides a graphic demonstration of rule categories and relationships as follows:

1. Simulates all the rule cluster transformations.

2. Distinguishes between symmetric, semi-asymmetric, and fully asymmetric rules.

3. Identifies special status rules resulting in collapsed clusters.

4. Accounts for the numbers of rules in equivalence classes and symmetry categories.

A4.1 Complimentary Transformations

R and R_c will be superimposed if any half division of the matrix is rotated over the other half (this is true of *all* numbering conventions).

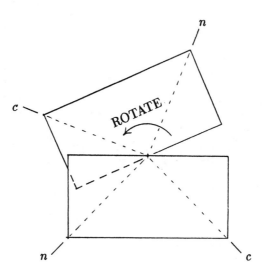

A4.2 Negative Transformations

Given a rule R, R_n is its reflection across the n diagonal. Alternatively, R and R_n will be superimposed if the matrix is *folded* across the n diagonal.

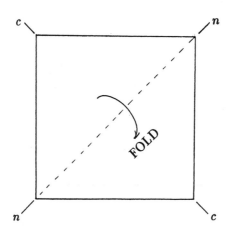

A4.3 Reflection Transformations

The matrix is shown subdivided into sixteen 4 × 4 segments, containing 16 rules. R and R_r pairs are contained within a segment, and are shown linked. Rules in each segment have the same characteristic layout according to their symmetry categories as illustrated below:

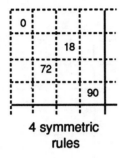
4 symmetric rules

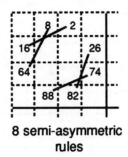
8 semi-asymmetric rules

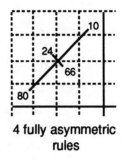
4 fully asymmetric rules

A4.4 Collapsed Clusters

The two diagonals across the matrix, labelled n (negative) and c (compliment), have special significance, because when the complimentary and negative manipulations of the matrix are carried out, rules related to the diagonals exhibit additional relationships, resulting in collapsed clusters as described in section 3.3.8.

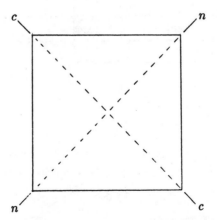

Rules that lie on the c diagonal have the property, for a given rule R, $R_c = R_n$. Fully asymmetric rules whose reflection link is bisected by the c diagonal, have the property, for a given rule R, $R_c = R_{nr}$. As stated earlier, R and R_n will be superimposed if the matrix is folded across the n diagonal; rules that lie on this diagonal (on the fold) have the property, $R = R_n$. Fully asymmetric rules whose reflection link is bisected by the n diagonal have the property that, for a given rule R, $R_n = R_r$.

Semi-asymmetric rules are unrelated to the diagonals in the ways described above, and therefore have no collapsed clusters.

240   APPENDIX 4 The Rule-Space Matrix n=3 Rules

A4.5 Symmetric Rules

The equivalence classes among the symmetric rules (total 36) consist of the rules which are superimposed when the matrix is folded across the n diagonal.

The 32 rules that have sometimes been designated as "legal"[33] (both symmetric and even) are located in the left half of the matrix. The 16 totalistic rules among the $n = 3$ rules are circled.

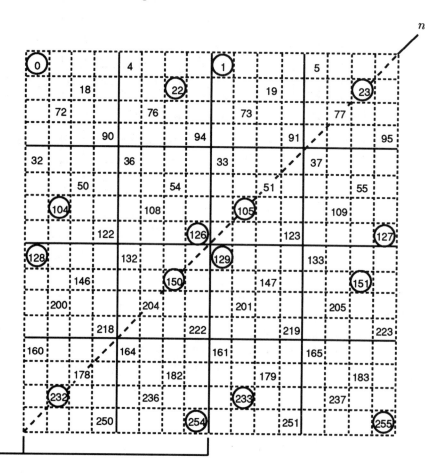

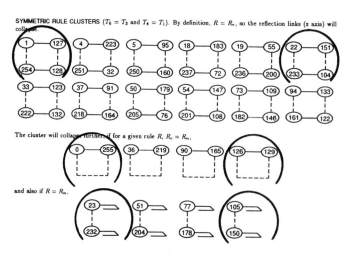

A4.6 Semi-Asymmetric Rules

The equivalence classes among the fully asymmetric rules (total 32) consist of the linked rules which are superimposed when the matrix is folded across the n diagonal.

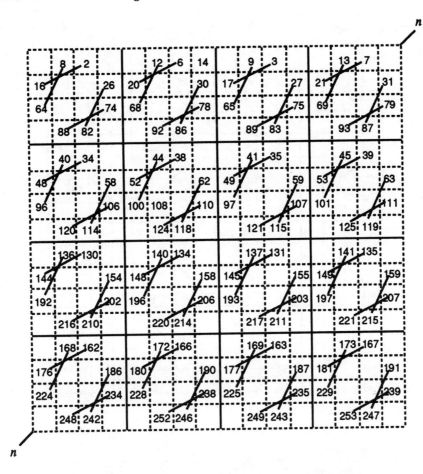

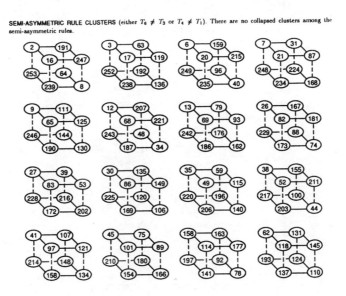

SEMI-ASYMMETRIC RULE CLUSTERS (either $T_6 \neq T_3$ or $T_4 \neq T_1$). There are no collapsed clusters among the semi-asymmetric rules.

A4.7 Fully Asymmetric Rules

The equivalence classes among the fully asymmetric rules (total 20) consist of the linked rules which are superimposed when the matrix is folded across the n diagonal.

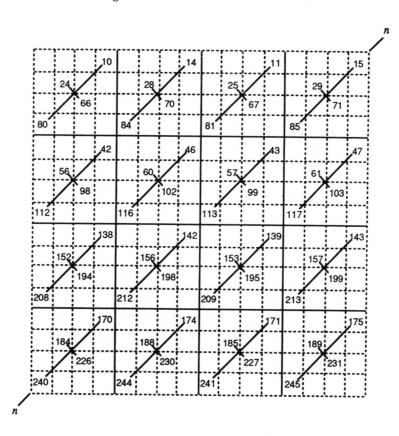

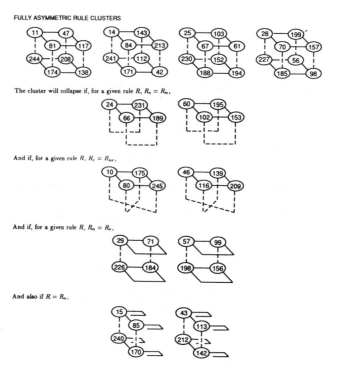

FULLY ASYMMETRIC RULE CLUSTERS

The cluster will collapse if, for a given rule R, $R_c = R_n$,

And if, for a given rule R, $R_c = R_{nr}$,

And if, for a given rule R, $R_n = R_r$,

And also if $R = R_n$,

References

1. Ashby, W. R. *Design for a Brain*. London: Chapman & Hall, 1960.
2. Berlekamp, E., J. H. Conway, and R. Guy. *Winning Ways for Your Mathamatical Plays*. New York: Academic Press, 1982.
3. Burks, A. W. *Essays on Cellular Automata*. Urbana: University of Illinois Press, 1966.
4. Dawkins, R. "The Evolution of Evolvability." In *Artificial Life*, edited by C. Langton, 201–220. Santa Fe Institute Studies in the Science of Complexity, Proc. Vol. VI. Redwood City, CA: Addison-Wesley, 1988.
5. Feldberg, R., and S. Rasmussen. Private communication, August 1991.
6. Grassberger, P. "Chaos and Diffusion in Deterministic Cellular Automata." *Physica D* **10D** (1984): 52–58.
7. Hogg, T., and B. A. Huberman. "Attractors on Finite Sets: The Dissipative Dynamics of Computing Structures." *Physical Review A* **32(4)** (1985): 2338–2346.
8. Jen, E. "Global Properties of Cellular Automata." *J.Stat.Phys.* **43** (1986): 219.
9. Jen, E. "Scaling of Preimages in Cellular Automata." *Complex Systems* **1** (1987): 1045–1062.
10. Jen, E. "Cylindrical Cellular Automata." *Com. Math. Phys.* **188** (1988): 569–590.
11. Jen, E. "Linear Cellular Automata and Recurring Sequences in Finite Fields." *Com. Math. Phys.* **119** (1988): 13–28.
12. Jen, E. "Enumeration of Preimages of Cellular Automata." *Complex Systems* **3** (1989): 421–456.
13. Kaneko, K. "Attractors, Basin Structures and Information Processing in Cellular Automata." In *Theory and Application of Cellular Automata*, 367–399. Singapore: World Scientific, 1986.
14. Kauffman, S. A. "Emergent Properties in Random Complex Automata." *Physica D* **10D** (1984): 146–156.
15. Kauffman, S. A. "Requirements for Evolvability in Complex Systems: Orderly Dynamics and Frozen Components." In *Complexity, Entropy and the Physics of Information*, edited by W. H. Zurek, 151–192. Santa Fe Institute Studies in the Sciences of Complexity, Proc. Vol. VIII. Redwood City, CA: Addison-Wesley, 1990.
16. Langton, C. G. "Studying Artificial Life with Cellular Automata." *Physica D.* **22** (1986): 120–149.
17. Langton, C. G. "Computation at the Edge of Chaos: Phase Transitions and Emergent Computation." *Physica D* **42** (1990): 12–37.
18. Li, W. "Complex Patterns Generated by Next Nearest Neighbors Cellular Automata." *Comput. & Graphics* **13(4)** (1989): 531–537.
19. Li, W., N. H. Packard, and C. Langton. "Transition Phenomena in Cellular Automata Rule Space." *Physica D* **45** (1990): 77–94.

20. Li, W., and N. H. Packard. "Structure of the Elementary Cellular Automata Rule Space." *Complex Systems* **4** (1990): 281–297.
21. Li, W. "Phenomenology of Non-Local Cellular Automata." Working Paper 91-01-001, Santa Fe Institute, Santa Fe, NM, January 1991.
22. Martin, O., A. M. Odlyzko, and S. Wolfram. "Algebraic Properties of Cellular Automata." *Comm. Math. Phys.* **93** (1984): 219–258.
23. Moore, E. F. "Machine Models of Self Reproduction." *Proc. Symp. Appl. Math.* **14** (1962): 17.
24. Packard, N. H., and S. Wolfram. "Two-Dimensional Cellular Automata." *J. Stat. Phys.* **38(5/6)** (1985).
25. Pitsianis, S., G. L. Bleris, Ph. Tsalides, A. Thanailakis, and H. C. Card. "Algebraic Theory of Bounded One-Dimensional Cellular Automata." *Complex Systems* **3** (1989): 209–227.
26. Von Neumann, J. "Theory of Self-Reproducing Automata." *1949 Univ. of Illinois Lectures on the Theory and Organization of Complicated Automata*, edited by A. W. Burks. Urbana: University of Illinois Press, 1966.
27. Walker, C. C., and W. R. Ashby. "On the Temporal Characteristics of Behavior in Certain Complex Systems." *Kybernetik* **3** (1966): 100–108.
28. Walker, C. C. "Behavior of a Class of Complex Systems: The Effect of System Size on Properties of Terminal Cycles." *Cybernetics* (1971): 1, 4, 55–67.
29. Walker, C. C., and A. A. Aadryan. "Amount of Computation Preceding Externally Detectable Steady-State Behavior in a Class of Complex Systems." *Bio-Medical Computing* **2** (1971): 85–94.
30. Walker, C. C. "Predicting the Behavioral Effects of System Size in a Family of Complex Abstract Systems." *Progress in Cybernetics and Systems Research* **111** (1978): 43–47.
31. Walker, C. C. "Stability of Equilibrial States and Limit Cycles in Sparsely Connected, Structurally Complex Boolian Nets." *Complex Systems* **1** (1987): 1063–1086.
32. Walker, C. C. "Attractor Dominance Patterns in Random-Structure Cellular Automata." *Physica D* (1990): in press.
33. Wolfram, S. "Statistical Mechanics of Cellular Automata." *Revs. Modern Physics* **55(3)** (1983): 601–64.
34. Wolfram, S. "Universality and Complexity in Cellular Automata." *Physica D* **10D** (1984): 1–35.
35. Wolfram, S. "Computation Theory of Cellular Automata." *Comm. Maths. Phys.* **96** (1984): 15–57.
36. Wolfram, S. "Twenty Problems in the Theory of Cellular Automata." *Physica Scripta* **T9** (1985): 170–183.
37. Wolfram, S. "Undecidability and Intractability in Theoretical Physics." *Phys. Rev. Lett.* **54(8)** (1985): 735–738.
38. Wolfram, S. "Approaches to Complexity Engineering." In *Theory and Application of Cellular Automata*, edited by S. Wolfram, 400–413. Singapore: World Scientific, 1986.
39. Wolfram, S. "Appendix: Properties of the k=2, r=1 Cellular Automata," Table 1-16. In *Theory and Application of Cellular Automata*, edited by S. Wolfram, 513–536. Singapore: World Scientific, 1986.
40. Wolfram, S. "Random Sequence Generation by Cellular Automata." *Advances in Applied Math.* **7** (1986): 123–169.
41. Wooters, W. K., and C. Langton. "Is There a Sharp Phase Transition for Deterministic Cellular Automata?" Part of the Proceedings of the 1989 Cellular Automata Workshop. *Physica D* **45** (1990): 95–104.
42. Wuensche, A., and M. J. Lesser. "Atlas of Networks of Attraction in One-Dimensional Cellular Automata." Paper presented at the Santa Fe Institute and IERC, Cranfield Institute of Technology, March 1990.
43. Warrell, G. J. Private communication, January 1992.

Index

Index

A

additive rules, 17, 28-29
ambiguous permutation, 29, 38
architecture, 5
    disordered, 5, 15-16
    local, 5, 16, 18
    non-local, 13
    ordered, 5, 16
array, 5-6
array length, 6, 10, 28
artificial life, 14
Ashby, W.R., 15
asymmetric neighbourhoods, 20
Atlas of basin of attraction fields, 13
    index of $n = 3$ rules, 83
    index of totalistic rules, 185
Atlas program, 61
attractor, 7, 9
    cyclic, 7, 10
    point, 10, 15
    strange, 51
attractor cycle, 9-10, 15, 17
    length of, 16
    maximum period, 17, 54
    trees rooted on, 8
attractor node, 9
attractor period, 10, 17

B

basin field, 11
basin (field) topology, 13, 21, 29, 54, 57
    in relation to rule class, 55
basin of attraction, 8-9
basin of attraction field, 8-9, 11, 15, 81
    atlas of
        see Atlas
    construction of, 11, 27
    significance of, 13
behaviour
    dynamical, 5, 7, 13, 16, 51
    emergent, 5
    equivalent, 19, 21
    global, 8, 11, 13
behaviour space
    topology of, 8
bilateral symmetry, 18, 21, 27
binary rule table, 56
binary value range, 6, 18
biomorphs, 14
boundary conditions, 5, 16

C

CA
    see cellular automata

cell, 5
cell's value, 5
cellular automata, 5
    evolution of, 6, 15-17
    finite, 8
    parameters, 15-16, 18
cellular automaton transition function, 8
chaotic space-time patterns, 51, 55
circular array, 6
clusters
    collapsed, 22, 239
    fully asymmetric rule, 23
    limited pre-image rule, 32
    rule, 18, 22, 24, 26
    semi-asymmetric rule, 23
    symmetric code, 27
    symmetric rule, 22
    totalistic code, 27
code
    see totalistic code, 26
collapsed clusters, 22, 239
complementary neighbourhood pairs, 19, 25
complementary transformation
    $n = 3$, 19, 238
    $n = 5$, 25
complex rules, 51
complex space-time patterns, 53
computing pre-images, 12, 38
continuous dynamical system, 8, 51
contraction map, 8
convergence of state space, 39, 52
Conway, J.H., 5
corrected Z parameter
    $n = 3$, 40
    $n = 5$, 44
cyclic attractor, 7, 10

D

Dawkins, Richard, 14
degree of ambiguity, 33
degree of pre-imaging, 9, 52
density of garden-of-Eden nodes, 39, 52
deterministic k-set, 32
deterministic permutation, 28, 38
    $K > 2$, 32
deterministic structure, 28
    $n = 1$, 34
    $n = 2$, 35
    $n = 3$, 36
    $n = 5$, 39
    one-way, 29, 37
    two-way, 29, 32, 35
deterministic template, 41
directed arcs, 8
disclosure length, 9, 15
    maximum, 17, 21

Index

discrete dynamical system, 5
disordered architecture, 5, 15-16
disordered state, 17-18
DNA, 13, 56
dominance, 16
dynamical behaviour, 5, 7, 13, 16, 51
dynamical system, 5, 13
    continuous, 51

E

elementary rules, 5-6, 18
emergent behaviour, 5
emergent structures
    interacting, 6
equivalence class, 21, 24, 26-27
equivalent basin, 10
equivalent behaviour, 19, 21
equivalent codes, 27
equivalent pre-images, 10
equivalent rules, 21
equivalent space-time patterns, 21
equivalent transient trees, 10
equivalents by interchanging cell values, 15
equivalents by reflection, 15
evolution
    of the cellular automata, 6, 17-18
evolutionary location, 8
excluded permutation, 29, 37-38
exhaustive testing, 11, 28

F

field, 8
finite cellular automata, 8
fully asymmetric rule clusters, 23, 158
fully asymmetric rules, 20, 24, 26, 242

G

game of life, 5
garden-of-Eden, 9
    density of, 39, 52, 54
    nodes, 9-10
    state, 7, 9-10, 28
genetic code, 14
genetic systems, 16
genotype, 14, 56
global behaviour, 8, 13
global state, 5
graphic convention, 9, 73

H

Hamming distance, 13, 56

hidden deterministic permutations, 40
historical time reference, 15

I

in degree, 9
incoming arc, 8
information structures, 51
initial condition, 5
initial global state, 6
input line, 19-20
    negative, 19
interacting emergent structures, 6

J

Jen, Erica, 28

K

Kauffman, Stuart A., 5, 13, 16, 18, 54, 56
K-set, 32

L

λ parameter, 51-52
    tables, 48-49
λ ratio, 52
    tables, 48-49
Langton, Christopher, 16, 51
left ambiguous permutation, 29
left deterministic k-set, 32
left deterministic permutation, 28
left deterministic structure, 28
left excluded permutation, 29
left start string, 28
legal rules, 240
length of attractor cycles, 16
limit cycles, 51
limit points, 51
limited pre-image rule clusters, 32
limited pre-image rules, 10, 27-28, 33
    in general, 28
    $n = 1$, 34
    $n = 2$, 35
    $n = 3$, 30, 36
    $n = 5$, 32, 39
    one-way, 29
    two-way, 29
local architecture, 5, 16, 18
local neighbourhood, 5-6
logical universe, 5

M

Martin, O., 8, 17, 28
mated rules, 59
maximum attractor cycle period, 17, 54
maximum disclosure length, 17, 21
maximum length of transient trees, 54
maximum pre-imaging, 29, 37, 39, 52, 54
mirror-image space-time patterns, 15, 20
mutants, 14, 59, 225
mutant basins of attraction, 56
mutation, 14, 56, 59, 225

N

NAT
    see network of attraction
nearest-neighbour wiring, 16, 18
negative input line, 19
negative space-time pattern, 19
negative transformation
    $n = 3$, 19, 238
    $n = 5$, 26
    $n = 5$ totalistic code, 27
neighbourhood, 5
    asymmetric, 20
    complementary pairs, 25
    local, 5-6
    $n = 3$, 19
    $n = 5$, 25
    $n = 5$ totalistic, 26
    template, 5
network of attraction, 8, 62
networks of Boolean functions, 5, 15
neural networks, 13
nodes, 8
non-local architecture, 13
number of separate basins in the field, 54
number of theoretic properties of the array length, 16, 18
numbered nodes, 11

O

one-bit mutant, 57, 225
one-way deterministic structure, 29, 37
one-way limited pre-image rule, 29
ordered architecture, 5, 16
ordered wiring, 5, 17
out degree, 9
outgoing arc, 8

P

parallel processing, 5

parameters
    λ, 48-49, 51-52
    of cellular automata, 15-16, 18
    Z, 39, 48-49, 52, 54
partial pre-image, 30, 38
    queue, 38
period
    attractor, 10, 17
    maximum attractor cycle, 17, 54
periodic boundary conditions, 6, 16, 25
    $n = 3$, 19
    $n = 5$, 25
permutation
    ambiguous, 29, 38
    deterministic, 28, 38, 32
    excluded, 29, 37-38
    hidden deterministic, 40
    left ambiguous, 29
    right ambiguous, 29
    right deterministic, 28
    right excluded, 29
phase portrait, 8
phase space, 7
phase transition, 16, 51, 55
phenotype, 14, 56
point attractor, 10, 15
pre-images, 8, 10-11, 15, 28
    computing, 12
    equivalent, 10
    of any rule, $n = 3$, 37
    of any rule, $n = 5$, 39
    of limited pre-image rules, $n = 2$, 34
    of limited pre-image rules, $n = 3$, 36
    of limited pre-image rules, $n = 5$, 38
    partial, 30, 38
pre-imaging
    maximum, 29, 37, 39, 52, 54

Q

quiescent (non-quiescent) rule table entries, 51

R

random Boolean networks, 5, 16
reflected input line, 20
reflected (mirror-image) space-time pattern, 20
reflected neighbourhood pairs, 25
reflection transformation
    $n = 3$, 20, 239
    $n = 5$, 26
repeat state, 9
repeating segments, 10, 16
reproduction, 14
reverse algorithm, 10, 12, 28, 33, 38
right ambiguous permutation, 29

Index

right deterministic k-set, 32
right deterministic permutation, 28
right excluded permutation, 29
right start string, 28-29
rotation equivalent states, 10, 17
rotation symmetry, 16-17, 21
rule algorithm, 33
rule class, 51
rule cluster, 18, 22, 24, 26
rule number, 19, 25
rule numbering system
    $n = 3$, 18
    $n = 5$, 25
    $n = 5$ totalistic code, 26
rule space, 13, 16, 51, 57
rule-space matrix, 24, 235
rule table, 13-14, 19-20, 28, 56
    binary, 56
    $n = 3$, 19
    $n = 5$, 25
rule transformation
    $n = 3$ to $n = 5$, 25
    totalistic code to $n = 5$, 26
rules
    additive, 17, 28-29
    complex, 51
    elementary, 5-6, 18
    equivalent, 21
    fully asymmetric, 20, 24, 26
    limited pre-image, 10, 27-28, 33
    mated, 59
    $n = 3$, 18
    $n = 5$, 25
    pre-images of, 34, 36-39
    semi-asymmetric, 20, 24, 26
    source, 57
    symmetric, 18, 20, 24, 26-27
    totalistic, 26, 240

S

seed, 9
segmented state, 17-18
self-reproduction, 5
semi-asymmetric rule clusters, 23, 125
semi-asymmetric rules, 20, 24, 26, 241
shift invariance, 10
significance of basin of attraction fields, 13
singleton state, 18
source rule, 57
space-time patterns, 6, 51
    chaotic, 51, 55
    complex, 53, 59
    equivalent, 21
    in relation to rule class, 55
    negative, 19
    reflected (mirror-image), 20

space-time trajectories, 15
start segment, 30
start string, 28-29, 38
state space, 7, 9-10, 18
    convergence of, 39, 52
state transition fragment, 8
state transition graph, 8-9, 15
    construction of, 9
states
    disordered, 17-18
    garden-of-Eden, 7, 9-10, 28
    global, 5
    initial global, 6
    repeat, 9
    rotation equivalent, 10
    segmented, 17-18
    singleton, 18
    successor, 9, 15, 27
    uniform, 17
strange attractor, 51
successor state, 9, 15, 27
suppressed equivalent transient branches, 10
suppressed equivalent transient trees, 10
symmetric code clusters, 27
symmetric rule clusters, 22, 84
symmetric rules, 18, 20, 24, 26-27, 240
symmetry categories, 20

T

target cell, 5, 15, 18, 25
templates
    deterministic, 41
    neighbourhood, 5
time step, 7-8
topology of basin of attraction fields, 16
topology of behaviour space, 8
totalistic code, 5, 26
    clusters, 27
    table, 26
totalistic rule
    $n = 3$, 240
    $n = 5$, 26
trajectory, 5
transient, 7, 9, 51
transient branch, 9-10
transient evolution, 17, 21
transient tree, 9-10
    construction of, 10
    maximum length of, 54
transition arcs, 9
transition function, 5, 15
translational invariance, 16
trees rooted on attractor cycles, 8
two-way deterministic structure, 29, 32, 35
two-way limited pre-image rules, 29

U
uniform state, 17
universal computation, 51

V
value range, 15
Von Neumann, J., 5

W
Walker, Crayton, 5, 8, 15, 18
wiring diagram, 5, 15
Wolfram, Stephen, 5, 8, 16, 19, 26, 51

Z
Z parameter, 39, 52, 54
    corrected, 40, 44
    tables, 48 -49

SEP 2 0 1993